Modern Cosmology
By: Paul Karl Hoiland
Email: joatp2000@yahoo.com

Modern Cosmology Introduction

Dedicated to all the curious out there.

Findings are pouring in, ideas are bubbling up, and research to test those ideas is simmering away. All of this has made for one of the most exciting times for modern Cosmologist. But there is one dark note in all this: All of the many different ideas cannot all be correct. All the ideas under discussion are not even consistent with one another. Before we attempt to make some sense of all this we must first take some of these ideas and compare them to understand the differences.

HOT BIG BANG STANDARD MODEL

In 1915 when Einstein was finishing the development of the General Theory of Relativity, the theory's prediction of an expanding universe was in conflict with firm philosophical beliefs of the day. Thus Einstein introduced a fudge factor, the cosmological constant, to force the universe to be static. When in 1929 Hubble showed that the universe really was expanding Einstein immediately dropped the cosmological constant, calling it "the biggest mistake of my life."

Up until the early 1960's, though, a vestige of the idea of a static universe remained in the steady-state model of cosmology. Championed by Sir Fred Hoyle and others, this model posited that although the universe was expanding, matter was being created everywhere in the universe so that the overall density of matter in the universe remains constant. The mechanism for the creation of this matter was never found.

As data began to accumulate on the large scale structure of the universe, the steady-state model began to be in increasing difficulty.

One of the most important of these appeared in 1965 when Penzias and Wilson at AT&T Bell Labs discovered an all pervasive isotropic microwave radiation, corresponding to what would be emitted by a body with a temperature of -270 C. This radiation is widely interpreted to be a remnant of the big bang, and is usually called the cosmic microwave background radiation.

After Penzias and Wilson's work, the Standard Hot Big Bang model described here "carried the day" and little more was heard from the proponents of the steady-state model. There was a big bang some 15 billion years ago, when the size of the universe was zero and the temperature was infinite. The universe then started expanding at near light speed. The sequence of events in this model is:

Time $t = 0$ (about 15 billion years ago)

Radius $r = 0$.

Temperature $T = $ Infinite.

Density = mass per volume = Infinite.

t = 0.01 seconds

T = 100,0.00,000,000 0C.

Energy is mostly radiation.

t = 2 seconds

T = 10,000,000,000 0C.

Density = 100 million kg per cubic meter.

Proton-antiproton and neutron-antineutron pairs begin forming.

t = 3 minutes

T = 1,000,000,000 0C.

Protons and neutrons begin forming hydrogen and helium.

t = 20 minutes

About 25% of the protons and neutrons in the universe are now helium.

t = 10,000 years

T = 10,000 0C.

Density = 0.000,000,000,000,000,01 kg per cubic meter.

Most energy is now mass, not radiation.

Condensation into stars begins. A photograph from the Hubble space telescope of a birthplace of stars appears below.

t = 15 billion years (now)

T = -270 0C. (This temperature is from the Penzias and Wilson experiment described above.)

Density = 10-27 kg per cubic meter.

In the Standard Hot Big Bang model, each part of the mass-energy of the universe is gravitationally attracted by all the other mass-energy of the universe, so the rate of expansion is expected to be slowing down.

A crucial question was whether this decrease in the rate of expansion is sufficiently great that at some point the expansion would stop and reverse.

If yes, then at some point in the future the size of the universe will again be zero with infinite density and temperature, the Big Crunch. We call such a universe closed. In this case the geometry of the space-time is similar to the surface of a sphere.

If no, the universe will expand forever. We call such a universe open. In this case the geometry is similar to the surface of a saddle.

A variation of Big Bang cosmology is called the Big Bang/Big Crunch. If the universe is closed it will end in a Big Crunch. But the conditions of the Big Crunch are identical to the conditions of the Big Bang. Thus the end of this cycle of the universe is the beginning of the next.

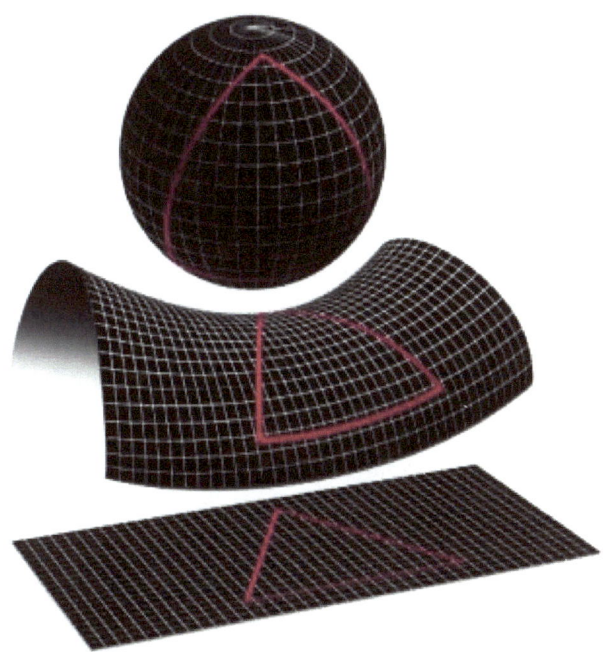

One problem is the moment of the Big Bang presents problems for physicists. The problem is that the language that we use to describe the universe, i.e. mathematics, breaks down when things become zero and when an infinity is reached. These conditions are called singularities. The mathematics works fine for any time after the Big Bang but not for moment of the bang itself. So the very expansion of the universe, well supported now by observational evidence in itself tends to impose some limits on our possible knowledge of creation unless we can find a way around this problem. In mathematics if we divide any finite quantity by zero the result is infinite. Another cases is zero divided by zero is undefined, by which we mean you can't do it.

Up until about a decade ago most people accepted the Standard Hot Big Bang model of cosmology. It accounted for a great deal of data regarding the large scale structure of the universe. Some problems were known, and then the list started to grow. These included: The flatness problem: Why is the matter density of the universe so close to the unstable critical value between perpetual expansion and recollapse into a Big Crunch?

The horizon problem: Why does the universe look the same in all directions when it arises out of causally disconnected regions? This problem is most acute for the very smooth cosmic microwave background radiation.

The dark matter problem: Of what stuff is the Universe predominantly made? Analysis of the gravitational interactions of galaxies shows much more matter than we can see. Nucleosynthesis calculations suggest that this dark matter of the Universe does not consist of ordinary matter - neutrons and protons?

Then in 1998 Perlmutter et al. published data that showed that, contrary to expectations, the rate of expansion of the universe is actually increasing. They measured the brightness and red shift of supernovae. The brightness is a direct measure of their distance away from us, and the red shift measures the speed of the supernovae away from us. Thus Perlmutter was taking the same sort of data as Hubble did 70 years before, but this time for supernova, which are much further away from us then the Cepheid variable stars that Hubble used.

HOT BIG BANG STANDARD MODEL WITH INFLATION

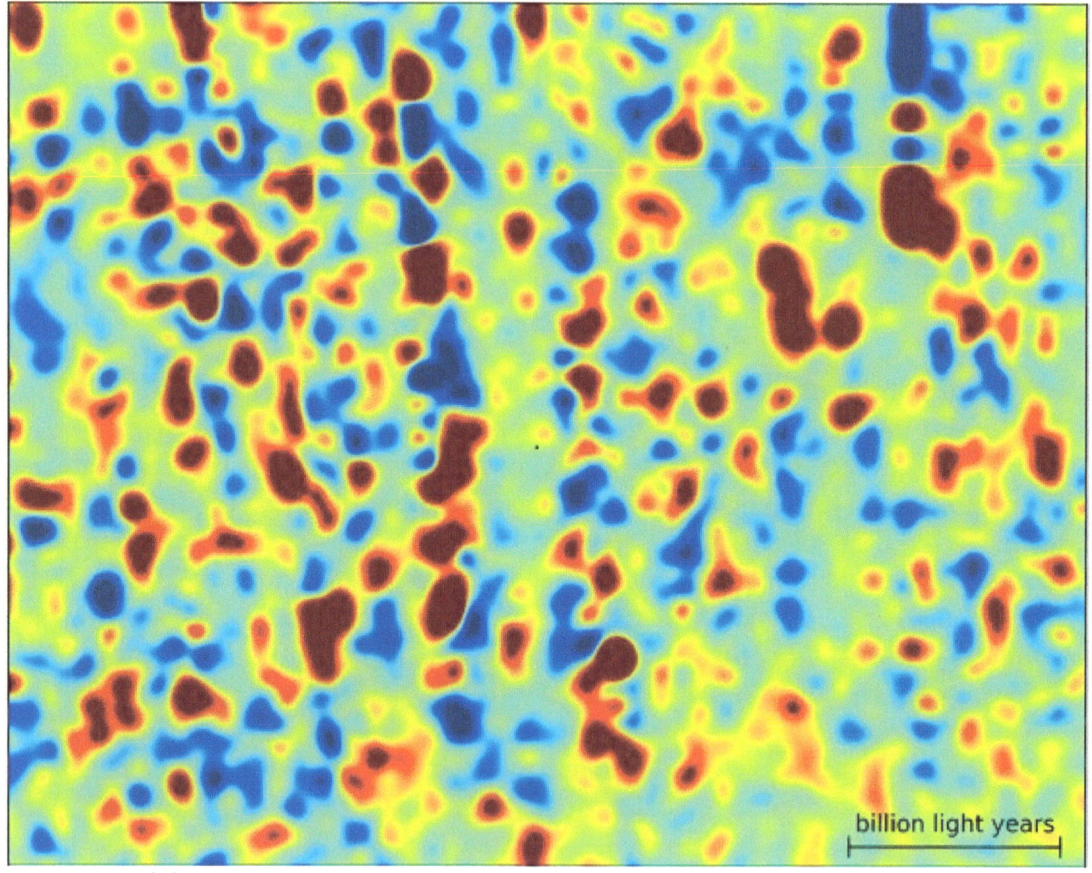

CMB map of the Cosmos in 3D

The early universe was nearly homogeneous, or the same in different places. Our most direct measure of this uniformity comes from observing the microwave background radiation that was emitted when the universe was roughly 300,000 years old. The intensity of this radiation is a direct measure of how dense the universe was at that time. Looking at this background radiation coming to us from different directions shows that the largest density differences from one point to another were about one part in 100,000. If the universe had been less homogeneous it would not have given rise to the smooth distribution of galaxies we see filling the sky. If it had been exactly homogeneous, however, then clumps of matter like galaxies would never have emerged at all. The big bang model offers no explanation for why the universe emerged in this nearly, but, not perfectly homogeneous state.

The second initial condition has to do with something called curvature. General relativity says that the universe can be closed, i.e. curved inward like the surface of a ball, open, meaning curved outward like the surface of a saddle, or flat, meaning it has no curvature. These different kinds of curvature cause the universe to evolve in different ways; a closed universe will eventually stop expanding and recollapse, while an open universe will tend to fly apart more and more quickly. For the big bang model to work the universe at the time of Planck density must have been almost precisely flat; the curvature couldn't have exceeded one part in 10^{59}.

Another set of problems with the big bang model has to do with the production of exotic particles at high energies. According to our current physical theories we believe that in the hot, dense environment prevalent in the early universe a number of exotic particles would have been produced. The current universe is far too cold to produce the reactions required to make these particles, but if they had been produced in the early universe we would expect some of them to still be detectable today. Although these particles could only have been produced in the first very small fraction of a second after Planck density we would nonetheless expect so many of them to have been produced that they would be quite abundant today. Any particle left over from the early, hot stages of the universe is called a relic particle. The big bang model predicts that we should see such relics, but we don't.

A solution to this problem came in the form of Inflation. The basic idea of inflation has to do with the rate at which the universe is expanding. When I use the term "rate" in this context I don't mean a speed. In an expanding universe the distances between galaxies are increasing, and the rate of expansion essentially refers to how long it takes for all of those distances to double.

In the standard big bang model the universe experiences power law expansion, meaning the doubling time gets longer as the universe expands. For example, in our current power law expansion distances in the universe were roughly half their current value about 10 billion years ago, but they won't be twice their current value until about 30 billion years from now. By contrast, if the doubling time stays constant then the expansion is referred to as exponential. Inflationary theory says that before our current power law expansion there was a brief period of exponential expansion.

Exponential growth can be much faster than power-law growth. In the simplest models of inflation the universe would have expanded by a factor of over ten to the ten million in a fraction of a second. Basically, our universe expanded faster than light which does not violate special relativity at all since it is space itself that is increasing in size.
In general relativity the rate at which the universe expands depends on the average energy density in the universe. If the density is high the expansion is rapid and the doubling time is small. The actual relation is that the doubling time is proportional to one over the square root of the energy density.

In general the expansion rate slows down as the universe expands because the average density decreases. If there are 1000 galaxies in some region of space and all distances double then the volume of space occupied by those galaxies will increase eight times. Since the galaxies have the same total mass as before their density will decrease by eight times. If the mass of galaxies were the only form of energy in the universe then every time distances doubled the doubling time would increase by a factor of the square root of eight. In short a universe whose energy consists entirely of mass will experience power law expansion.

Inflation doesn't require precise exponential expansion. Rather there is a set of mathematical criteria for how close to exponential the expansion needs to be during inflation, i.e. how much the doubling time can change each time distances double, in order for inflation to still have the consequences described below. Given a scalar field with a high enough energy density these conditions will be met and the expansion of the universe can be considered quasi-exponential. In general, however, the energy density of a scalar field is not perfectly constant as the universe expands. Rather it decreases more rapidly the smaller it is, such that eventually when it becomes small enough the universe enters a stage of power-law expansion.

So in order for inflation to have occurred it suffices that some scalar field exists and at some point in the past it had a very large energy density. While it is true that we have never to date observed a scalar field, physicists believe for a variety of theoretical reasons that many of them probably do exist and that we will start to see them in our next generation of particle accelerators. We could then ask why there was a scalar field? Recall that the "initial conditions" for our universe were set by physics that we don't know occurring above the Planck scale that we cannot study directly yet.

All we require for inflation is that somewhere there was one region, no matter how small, where the largest contribution to the energy came from a high-energy scalar field. If that happened then that small region would inflate, almost instantly growing much larger than all the other regions around it. Very soon this inflationary region would occupy nearly 100% of the total volume.

At the same time inflation not only answers the objection of why our universe appears so homogeneous. It also supplies an answer of what happened to the relic defects and particles like monopoles.

VARIABLE SPEED OF LIGHT MODEL

Basically, this variant on modern cosmology starts with the same basic Big Bang conditions with or without inflation except there is one major difference. That difference is found in the speed of light no longer being a constant.

In the seventies, two of Lebedev's physicists, A. D. Linde and D. A. Kirzhnits, using Fradkin's formalism, proved how with the decrease of temperature the universe passes through a phase transition that produces the breaking of the unified electroweak force into the electromagnetic force and the weak-nuclear force. But Kirzhnits was also the first to show that a particle possessing the tensor mass such that m1 < m0 can travel superluminally. Today we call these particles tachyons. However, VSL is based upon a concept a bit different from the older theories of tachyons. With VSL you still have regular C limited particles possessing normal mass. It is the velocity of light itself that is seen to change either over time, or scale, or both.

Basically, under these models, and there are several that have been proposed, C has varied over time either due to a changing vacuum state, due to brane world conditions, or to some other exotic solution. Popularized in modern times by such men as Lee Smolin, Joao Magueijo, Myself, and others this addition to the Big Bang model has some modern observational evidence in its favor. However, it remains outside the mainstream model employed and does not in itself attempt to over throw the basic premise of Special Relativity. Basically, this model solves one major problem which deals with the transfer of information across the cosmos and relates to the original Big Bang horizon problem.

Bell's Inequality and EPR Problems

The apparent contradiction in fact discloses only an inadequacy of the customary viewpoint of natural philosophy for a rational account of physical phenomena of the type with which we are concerned in quantum mechanics. Indeed the finite interaction between object and measuring agencies, conditioned by the very existence of the quantum of action, entails the necessity of a final renunciation of the classical ideal of causality, and a radical revision of our attitude towards the problem of physical reality.

-Bohm

The fact that Bell's Inequality is broken is considered by some to imply that one or both of these assumptions, locality or reality, must be false. But for any system where the light cone is expanded locality becomes an extended playing field and causality itself takes on extended meaning. Bohm's original pilot wave theory is but one of the different expanded light cone theories out there.

Some have proposed that the EPR problem must be seen as an ultimate attempt on the part of Einstein to prove the incompleteness of quantum mechanics, while circumventing the quantum postulate by measuring a physical quantity without interaction. But turning this around on itself, and from the perspective of modern experiments the EPR problem speaks to the incompleteness of quantum mechanics and to the completeness of Einstein's Special Theory of Relativity when it comes to a proper description of all information transfer within this universe. At the same time it also raises questions as to how far we can measure a physical quantity without interaction and how much we really understand cause and effect, so crucial an element to our comprehension of time itself.

The EPR problem has forced Bohr to change his interpretation from an interactional to a relational one. Thus, it is not only the interaction between the microscopic object and the measuring instrument (for particle 1) that is thought to be instrumental in defining the `element of physical reality' of particle 2, but also the correlation of the quantities of the two particles:

relation = interaction + correlation

Einstein realized that this renationalize introduced a feature of nonlocality into the Copenhagen interpretation which stood in opposition to his strict locality from Relativity. But this is only because of the defining in a certain way of equality of frames of reference. If, following modern conjecture, the specific frame of reference encountered in the EPR problem is a frame where time and space do break down, then there is no specific reason to invoke an equality of frames of reference which would then mean the EPR case circumvents the implications of relativity when it comes to the exactness of locality measurement.

It was Einstein's conviction that the EPR experiment can be understood in a local way if the state vector is not considered a description of an individual object but of an ensemble. Then the discontinuous change of the particle 2 state vector can be understood as a selection of a sub-ensemble, which does not seem to imply any real influence on particle 2 by the measurement of particle 1. Indeed, in many ways the modern concept of a quantum foam would find that Einstein's conviction was at least partly correct and that what we have at play here is sub-ensembles being selected together to form the end result. But this is only part of the solution to what is actually going on at the particle and sub-particle level.

When you examine such issues as brane lensing you encounter cases where from the brane perspective one can have actions, though limited to our normal understanding of C appear to be non-local to each other in one frame of reference, ie that of the local brane itself, but, which, from another frame, ie that of the more macro-scale, are actually connected by the same light cone state. Brane lensing in the reverse direction can generate a similar appearance of odd definitions of locality where the reverse would hold true. Yet, in each reference situation there is upon closer examination no actual altering of the local velocity of light from the norm. Which then begs the question whither there actually is a variable C or is the problem more one of our definition of frame of reference.

MEASUREMENT AND OBSERVATION

Fritjof Capra thinks, "the human consciousness plays an important role in the course of observation" (Capra 1982/1988, S. 90):
My conscious decision how I want to observe the electrons determines the characteristics of the electron to a certain degree. If I ask a particle-question
– I will get a particle-answer. If I ask a wave-question – I will get a wave answer.
(Capra 1982/1988, S. 91)

Unless, we modify this thought a bit, we are led to assume The electron has no characteristics, which are independent from our consciousness. The structures, which are observed by scientists in nature, are closely connected with the structures of the consciousness, with their conceptions, thoughts and values. (Capra 1982/1988, S. 91)

If our future action and our past action prior to the point of observation where the wave function collapses determines the outcome then our presence is seen to modify and alter space-time in such a way that we simply bring about the collapse of the wavefunction to the state our presence predetermined. Locality, is simply the result of an overlap of time displaced states. The reason these displaced states are seen as non-local is simply because part of the picture is unable to be seen due to our lightcone limit on information. The particle is real in the past and the future. The Particle has an existence at the overlap junction in both the past and the future. These two times sets are vital to a true quantum understanding. As these two time sets overlap they cause the collapse of the wavefunction into the same reality that was always present in the first place. We just simply view only part of the picture because one state takes place far faster, than our ability to observe it, in its off brane state.

The possibility for the "behaviour" of a quantum-system is given by the motion-equation (Schrödinger-equation) for the state-vector Ψ and the possibilities, which are realizing by chance. Before measuring we have only information about the discrete spectrum of eigenvalues of observables. We lack the future set of data to take into account. That data is hidden from us by our own lightcone restrictions. The result is we observe the collapse, without part of the data that brought about the collapse. If we could somehow cancel out the past there would be no collapse in our observation. With no collapse, you have no particle state. The same goes if we try to separate the future state-vector. However, we can alter the future or past state-vector and still bring about a collapse. This is a sort of reversible butterfly effect at the quantum level.

Quantum Theory shows us, that our world at a fundamental level is not imaginable as "mechanistically" sum or interaction of "really" isolated things. The mechanistical word-view assumes, that bodies are isolable, space and impulse would be infinitely exactly measurable, and the law of dynamic motion would be identical with causality. The mechanistical worldview is not necessarily identical with the science of quantum physics. Quantum actions show that these assumptions can fail. In essence a fully deterministic outlook on the whole world around us fails at the quantum level which indirectly effects things at larger scales. Objects and even actions have an inner structure, work permanently in interactions, are determined by more characteristics than space and impulse and have statistical laws governed by both our past and our future. There is a reality out there we simply have strong limits on our ability to measure. That lack of ability to measure puts limits on our ability to predict, to fully rule out certain possible aspects, and leaves a strong gap in our understanding of natural process, especially at the quantum level upon which everything is framed.

STRING & BRANE WORLD COSMOLOGY

String Cosmology deals mostly with pre-Big Bang conditions in an attempt to answer questions about the early energy condition of the universe.
Pre-big-bang scenario: the general picture

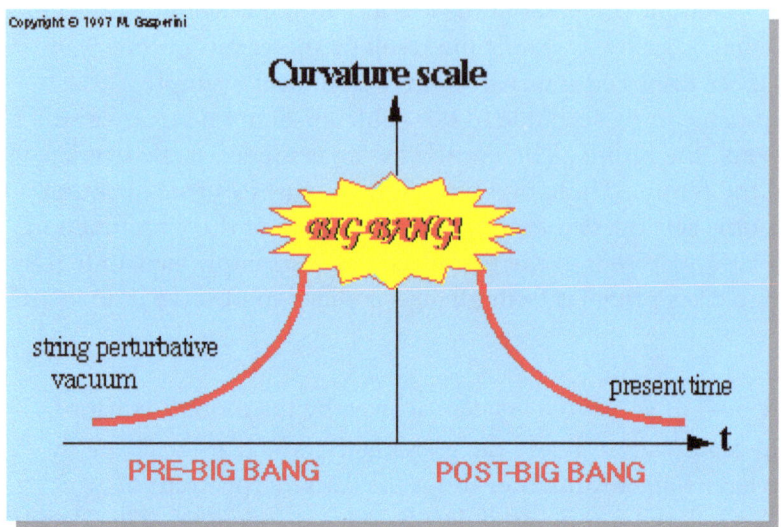

Basically, these models concern themselves with how our vacuum state initially got started. One of today's hottest subjects when it comes to Physics is M-Theory. But even in spite of String Theory solving a lot of problems(ie. Entropy of Black holes) it still in its current form commonly called M-Theory for Membrane or Magic has many problems of its own. String Theory from it's start has always been based upon another theory called Super Symmetry. With SUSY type theories all matter particles have their counter-part in force carriers. The idea being that Fermions can transform into Bosons and back. The problem is that nature, as we know it, has no observational evidence for the Supersymetry partners.

Another problem to some people is the added extra dimensions. These again have no direct observational evidence in nature, even though, some of the current modifications and extensions to the original String Theory have proposed experimental means of testing such.

Brane Cosmology deals with the same, but also goes a bit further in an attempt to explain not only what is outside of our Universe in what we termed hyperspace, but, also, where the Strings and membranes themselves come from. Both of these models employ dimensions beyond our standard model four of three spatio and one temporal from General Relativity. Into this mix I might also add Lee Smolin's much popularized Loop Quantum Gravity since under that model an attempt is made to determine what the Strings and Membranes themselves are formed from. Basically, both String Theory and LQFT are attempts to unite General Relativity with Quantum Field Theory.

LOOP IN TIME COSMOLOGY

First proposed by Richard Gott, this model basically follows the Big Bang Model with Inflation, except that it attempts to answer where the basic building blocks of space-time came from via a loop in time. For more information I would suggest reading Gott's book, "Time Travel within Einstein's Universe". The Model in some expanded versions shows up under Brane Cosmology with other versions of loops in time or cycle type creation models.

Basically, all these models are based upon modern physics theories. All of these models make an attempt to answer one of more problems raised by the study of cosmology, modern theory, modern observational evidence, or experimental evidence. While many of these models predict different results all of these models share a general background model basis we call the Standard Model of particle physics. They also all attempt to provide answers that fit the modern observational and experimental evidence to date. Basically, Modern Cosmology has come a long way from the old days of religious myth. But we modern Cosmologists share one thing in common with the creators of those old Myths: We have a wonder of the Cosmos within us and a desire to explain were we and everything around us came from. Today we rely on experimental and observational support for our theories. Yesterday man relied upon explanations that made sense to our limited knowledge we had at the time. Today our knowledge has increased manyfold to the point we can peer back in time across the Cosmos towards some of the earliest moments of creation and leep beyond even that with the power of mathematics and logic. Cosmology at its heart is about change. I have no doubt's that in time as we learn more even some of these many different modern models will themselves give way to even more interesting models.

We study an every changing and evolving process made up of many still untold processes. We peer back in time and attempt also to predict forward in time on grand time scales that are hard to fathom. But this is the job of a modern Cosmologist to study the great unknown and attempt as once only religions were allowed to do the answer to such questions as where did I come from and is this all there is?

SOME QUESTIONS

IS THE SPEED OF LIGHT CONSTANT?

No. The speed of light is different in different substances. For example the speed of light in water is 3/4 of the speed of light in a vacuum. This is usually expressed as the refractive index of water being 4/3.

Is the speed of light in a vacuum constant?

In some ways that depends on what you mean by constant. There are at least 3 possible meanings to this.

1. Is the speed of light the same in all directions at any one location? This is what is tested by the Michelson Morley experiment and the answer is "yes" for inertial frames of reference. For non-inertial frames of reference the answer is generally "no" as demonstrated by the Sagnac experiment.

2. Is the speed of light in vacuum the same at all times? Yes, because it is defined that way. It is defined to be 299,792,458 m/s. However if it was defined differently it might conceivably vary by ~1 part in 10^{10} per year but if it did so it would have to be related to variations in other "constants". It is also very possible it has varied over the time of the Cosmos in general and may actually be slowing down over time. It is also possible that exotic vacuum states allowed by quantum theory may have a different velocity of light within them. However, this being the case these exotic states do not mean the principles of Special Relativity are broken since its already known C is different in different mediums.
3. Is the speed of light in vacuum the same at all places? In terms of common conception the answer is no. Light is observed to bend when it travels very close to the sun. Such bending is due to a variation in the speed with distance near massive bodies.

4. There is another aspect to the non-constancy of the speed of e/m waves which is worth mentioning also. At the point of emission the speed of e/m waves are actually 1.732c (sqrt(3)c) because in the wave equation the time part of the wave is balanced against the 3 spatial dimensions at once. It is only after a wave has traveled at least one wavelength or more and can be approximated by a 1D plane wave that it slows down to c.

WHY IS VSL POPULAR IN SOME CIRCLES?

Part of the reason VSL has become popular in certain circles is because it answers the question of how observable casually disconnected regions of the cosmos can have communicated with each other. It also provides an answer to certain observations that seem to show we have photons with energies higher than those predicted under normal relativity and quantum theory. However, I'd be remise if I did not mention that those same observations have a still unsolved debate going as to whither or not the observational findings are correct. Also, some of us who hold to VSL have proposed that it also provides a solution to how possibly entanglement as observed in lab experiments might actually work in spite of casual disconnection of events. Here again it must be mentioned that this is not the only solution and much debate still exists on what type of information is actually being transferred in entanglement cases.

Generally, I myself have always favored a version of VSL that preserves the main implications of Special Relativity, that being Lorentz Invariance. But here again the debate and the models are wide open as recent articles on theoretical Lorentz Symmetry breaking that have appeared in modern Physics journals and in the research sharing archive systems like Lanl testify to. Basically, if Lorentz invariance is preserved then locally in any directly observable frame from our point of view Einstein stands as totally correct. So the version of VSL that I have tended to promote in no way debates the correctness of Einstein in general everyday situations.

DOES RELATIVITY OR THE MICHELSON MORLEY EXPERIMENT DISPROVE THE EXISTENCE OF AN ETHER?

1. No. What the M-M experiment proved was that matter was not a separate entity from the luminiferous (light carrying) ether but that matter was as bound to the ether as electromagnetic waves were. General Relativity is an ether theory as is the Zero Point field from QM. However, the ether in both is different from that employed by Newton. Our modern ether is the vacuum itself and this ether has no visual counterpart to an absolute frame of reference. We may say that according to the general theory of relativity space is endowed with physical qualities; in this sense, therefore, there exists an ether. As Einstein, in later comments put it, "the general theory of relativity space without ether is unthinkable". [From "Ether and the Theory of Relativity" an address delivered on May 5th, 1920, in the University of Leyden by Albert Einstein.]

What is the Standard Model?

"In particle physics, the description of the world in terms of fundamental particles and their interactions, which are mediated by the exchange of gauge bosons." [see Q is for Quantum by John Gribbin]. The Standard Model is a collection of sophisticated and respected theories of certain phenomena in nature: among them, what we call, the four forces of nature – the strong and weak nuclear forces, electromagnetism, and gravity. It has been exceedingly difficult to conceptually unify these forces within the conventional framework.

NOTES FOR REFERENCE

Defining a Cosmological Model
The requirements for a model
A cosmological model requires the following four basic concepts:
• A paradigm, ie a basic concept of how the Universe might be or function
• A distribution function for matter: a statement or mathematical expression for the way in

which mass (matter) is spread out in the Universe
• A theory of gravitation: usually based on the gravitational theories of Newton, Einstein and Mach
• A system of dynamics, these days based on relativity, giving the physics of the ways in which objects move in the Universe and the Universe expands.
A good cosmological model should:
Explain or take into account
• Red shift (Hubble's Law). It is observed that the light from distant objects has a longer wavelength (is red shifted) and correspondingly lower quantum energy, and that the red shift is roughly proportional to distance. This is generally taken as evidence for an expanding universe.
• The observed structure of the Universe (galaxies etc). Modern analysis of astronomical observations reveal a hierarchical clustered structure. Examples of levels of clustering are, in sequence, the Earth-Moon system, the Solar System, our (Milky Way) Galaxy, and the Local Super cluster of galaxies.
• The relative abundances of chemical elements, particularly the light ones (hydrogen, deuterium, helium, lithium).
• The existence of a dark sky (Olbers' Paradox), and the existence, temperature and isotropy of the cosmic microwave background radiation.
and make predictions which are
• Useful (otherwise the model is of no value)
• Falsifiable (ie able to be confirmed or refuted by astronomical observations).

Quantum Gravity

Under classical gravitation the gravitational field at a fixed time can be described by the geometry of the three spatial dimensions at that time. The history of the gravitational field is described by the four dimensional space-time that these three spatial dimensions sweep out in time. Therefore the path integral is a sum over all four dimensional space-time geometries that interpolate between the initial and final three dimensional geometries. In other words it is a sum over all four dimensional space-times with two three dimensional boundaries which match the initial and final conditions. The classical description of gravity is given by general relativity, which says that the gravitational force is related to the curvature of space-time itself. Unlike for non-gravitational physics, space-time is not just the arena in which physical processes take place but it is a dynamical field. Therefore a sum over histories of the gravitational field in quantum gravity is really a sum over possible geometries for space-time. However, we are now just beginning to learn that at a quantum scale space-time takes on a very strange landscape of gravity, which in some cases can begin to look like the movie Stargate on steriods with wormholes every where.

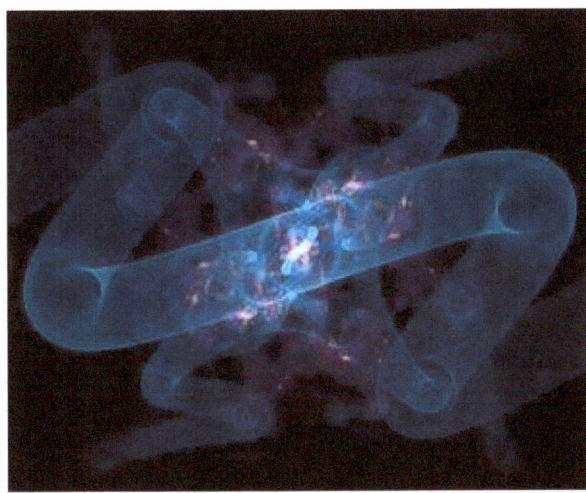

The path integral formulation of quantum gravity has many mathematical problems. It tends to break down at the Planck scale. However it can be used to correctly calculate quantities that can be calculated independently in other ways like black hole temperatures and entropies.

One way to look at it all is the Instanton viewpoint. Typical instantons resemble (four dimensional) surfaces of spheres with the three geometry slicing the sphere in half. They can be used to calculate the quantum process of universe creation, which cannot be described using classical general relativity. They only usually exist for small three geometries, corresponding to the creation of a small universe. Note that the concept of time does not arise in this process. Universe creation is not something that takes place inside some bigger space-time arena - the instanton describes the spontaneous appearance of a universe from literally nothing. Once the universe exists, quantum cosmology can be approximated by general relativity so time appears.

The first attempt to find an instanton that describes the creation of a universe within the context of the `no boundary' proposal was made by Stephen Hawking and Ian Moss. The Hawking-Moss instanton describes the creation of an eternally inflating universe with `closed' spatial three-geometries. Other attempts have dwelt with is the Universe flat, closed, or open.

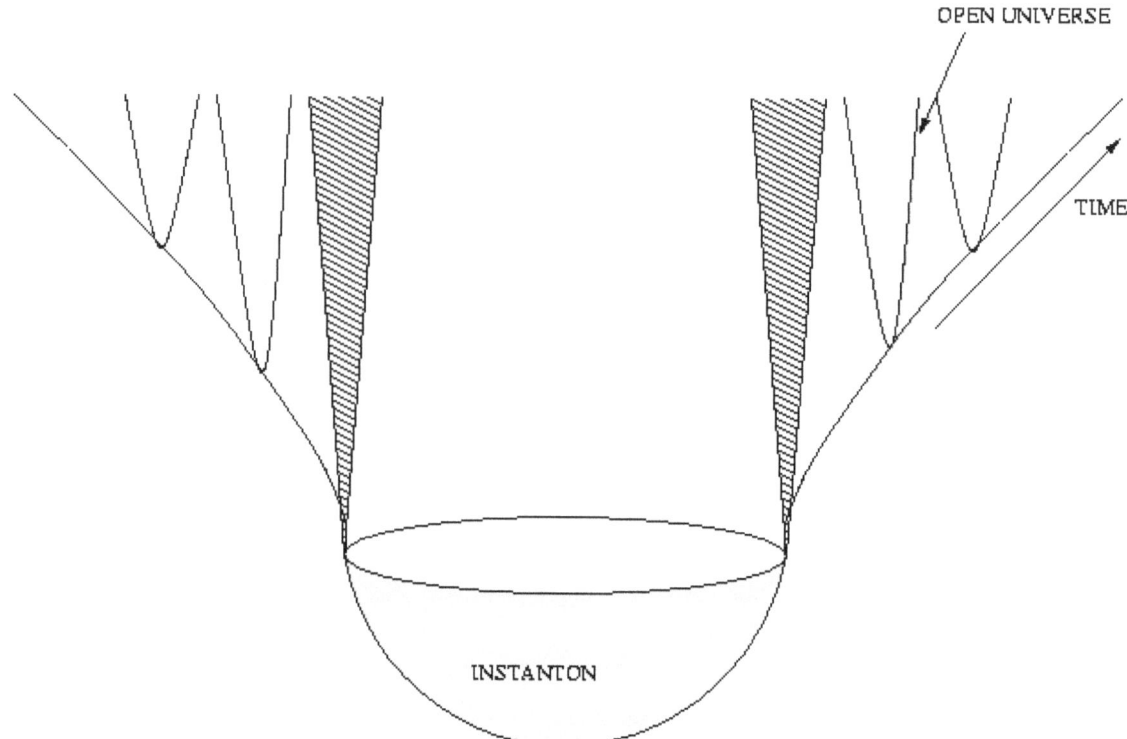

The Coleman-De Luccia Instanton

The idea behind the Coleman-De Luccia instanton, discovered in 1987, is that the matter in the early universe is initially in a state known as a false vacuum. A false vacuum is a classically stable excited state which is quantum mechanically unstable. In the quantum theory, matter which is in a false vacuum may `tunnel' to its true vacuum state. The quantum tunnelling of the matter in the early universe was described by Coleman and De Luccia. They showed that false vacuum decay proceeds via the nucleation of bubbles in the false vacuum. Inside each bubble the matter has tunnelled. Surprisingly, the interior of such a bubble is an infinite open universe in which inflation may occur. The cosmological instanton describing the creation of an open universe via this bubble nucleation is known as a Coleman-De Luccia instanton.

TACHYONS

Tachyons are associated with imaginary mass. Because

$$E = \frac{m_0 c^2}{\sqrt{1 - v^2/c^2}}$$

the denominator becomes imaginary if $v > c$. However, if the mass is observed as being imaginary, then E is real again. From a mathematical point of view Lorentz transformations are easily extended in Minkowski space to address velocities greater than c.

Imaginary mass sounds very much like an unphysical mathematical construct. But is this really so? For the Higgs mechanism we need just that, tachyonic mass. This is a fact rarely emphasized, i.e. the conceptual issues and the physical interpretation. The general interpretation since no physically detected version of a tachyon has been found is to consider them artifacts of the math involved. But this is an assumption that may not stand up to indirect evidence from Field theory itself.

If we introduce a scalar complex field with a potential, then only if we require $m^2 < 0$ do we get the famous Mexican hat potential making the vacuum degenerate and allowing for non-zero vacuum expectation values for all the particles. So already here we realize that our Higgs field and Lagrangian are not to be interpreted in a standard QFT manner. It appears as though the cornerstones of the standard model rely on unobservable features: scalars and tachyonic mass. When you extend this into String Theory with SUSY you again encounter Tachyons.

In 1974 bosonic string theory was plagued by tachyonic excitations of the ground state. The introduction of space-time SUSY by Wess and Zumino cured this problem with tachyon condensation. Recent string/M-theory developments re-address the tachyon issue within the context of D-branes. The tachyon can have a potential. The fact that it is a tachyon means that you're expanding about a maximum of the potential. However, the potential could have a minimum which would be a tachyon free vacuum. This is believed to be the case for a number of open string tachyons, but at the present is unproven by experimental evidence and only supported by theory.

The speed of propagation along a string itself is dependent upon the total energy stored in the String, P_{sub0} defined by

$C_{substring} = sqrt(fg)$,

where, f and g are functions of the total energy. Basically, the String tension is modified by the total energy stored and as such the tension is no longer a scalar, invariant for all observers which translates to the String's frame being different from the frame of the external observer. But even the external frames, while sharing an invariant speed of maximum propagation, are themselves variant when it comes to the separation distance between two frames due to the "warp factor" bringing about a brane equal to gravitational lensing. Thus, some of the better theories we have today(String Theory, M-Theory, Brane Theory, Loop Quantum Gravity, etc, as well as aspects of the Standard Model seem to imply we cannot always judge every action-reaction by what we in the macro-world can judge observationally which further implies our ideas of cause and effect need to be slightly modified.

The constancy of the speed of light is not a necessary condition for "evolutionary science" to be valid. This makes sense, in light of the fact that "evolution science" was strictly Newtonian until early this century, and Newtonian physics does not permit a constant speed of light, either cosmologically or otherwise. However, the constancy of the speed of light does have implications for Cosmology since certain prime aspects of our interpretation of observations are based upon the assumption that C is constant and that we can observationally compare frames. So no matter how the current VSL debate turns out it will have implications for both Physics and for Cosmology.

UNDERSTANDING UNCERTANITY(A TRIP INTO EXOTIC GEOMETRY)

The uncertainty relations from quantum mechanics, when interpreted as a physical axiom, prohibit any local and realistic interpretation, because their immediate consequence is a spreading of wave packets that rather runs against the whole construction of a conventional approach based upon relativity. But from a 5D perspective if we have action/reactions that can transpire in manifolds we are restricted in our ability to directly observe then it is possible to maintain a local and realistic interpretation simply because our definition of local becomes relative itself.

Since the relations are a cornerstone of quantum theory it seems that we have in the most modern theories a natural progression of quantum theory towards unification with the more conventional approaches. Whither or not such unification will ultimately expose new and unique methods of travel remains for theory and experimental evidence to show us. But the geometry of space-time has its own secrets that it is waiting to show us.

-Richard Feynman

"There was a time when the newspapers said that only twelve men understood the theory of relativity. I do not believe that there ever was such a time. ... On the other hand, I think it is safe to say that no one understands quantum mechanics. ... Do not keep saying to yourself, if you can possibly avoid it, `But how can it be like that?', because you will get `down the drain' into a blind alley from which nobody has yet escaped. Nobody knows how it can be like that."

I have pointed out in the past to someone who had a strict religious idealism and had problems with quantum mechanics that at some levels everything we accept as truth comes to us via senses that depend upon particles we cannot actual see or fully measure. In essence, her ability to read the Bible and even our ability to prove anything is real is itself dependent upon our observation of such. As I mentioned in the Incident at Chanute what our minds view as reality becomes reality for us. However, when you try to follow all of it out you do end up down a blind alley into a maze with no pathway out one can determine fully exists.

The idea we could be nothing more than some image in some giant holographic machine is just as valid as we are independent conscious beings of our own right when it comes to any quantum proof of existence. That applies to our understanding of everything around us, to our very lives and existence. Boil it all down, we cannot prove anything actually exists, except as some construct of a lot of interplay of energy at the quantum level. At some level we all revert to some aspect of an assumption based upon what we observe that such is real, when in fact, the aspect of reality is itself dependent upon observation of effects we cannot witness directly and can only approximate via math. Our macro-world view actually has its basis in a reality that is totally unseen and as strange as Alice in Wonderland's weird trip down the rabbit hole.

Perhaps in modern times we are beginning to see how it can be like that.

AN INTRODUCTION TO STRING THEORY

String theory is based on the idea that at Planckian scales, where the quantum effects of gravity are strong, particles are actually one-dimensional extended objects. Just as a particle that moves through space-time sweeps out a curve (the worldline)

string will sweep out a surface (the world-sheet)

interactions, supersymmetries and gauge groups. In fact, all the usual particles emerge as excitations of the string and the interactions are simply given by the geometric splitting and joining of these strings:

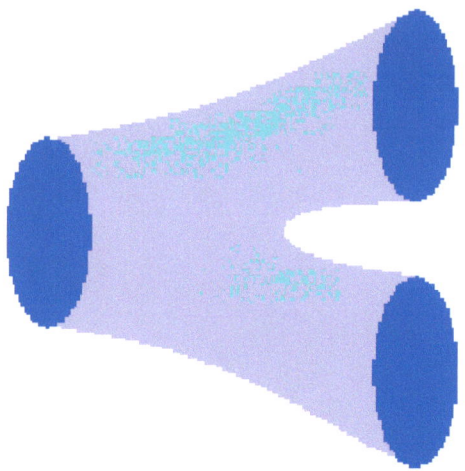

In this way the usual Feynman diagrams of quantum field theory are generalized by arbitrary Riemann surfaces

Much recent interest has been focused on D-branes. A D-brane is a submanifold of space-time with the property that strings can end or begin on it.

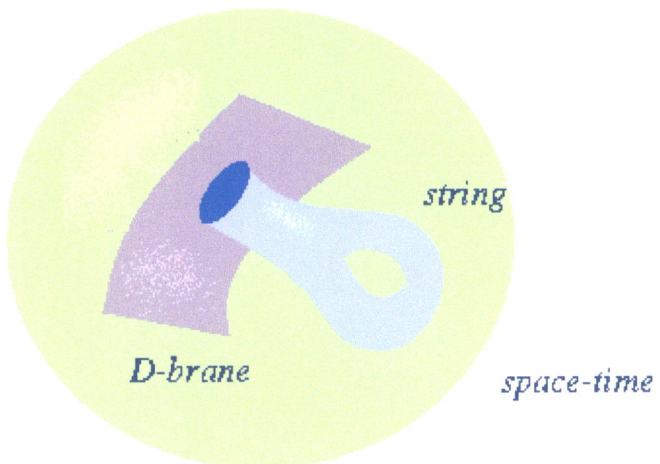

string

D-brane

space-time

If you are trying to build a model of our universe using Superstring Theory or M-theory, possibly curling up the higher dimensions to get our four, the Mathematics of the higher dimensions in which these theories like to live (10 or 11) can be communicated to the four dimensional physics (e.g. the observed particle spectrum) in a way which is mediated by the types of D-branes that can wrap (part of) themselves on the internal geometry and appear as particles or strings in the four dimensions.

Open strings can have two different kinds of boundary conditions called *Neumann* and *Dirichlet* boundary conditions. With Neumann boundary conditions the endpoint is free to move about but no momentum flows out. With Dirichlet boundary conditions the endpoint is fixed to move only on some manifold. This manifold is called a ***D-brane*** or ***Dp-brane*** ('p' is an integer which is the number of spatial dimensions of the manifold).

For example we see open strings with one or both endpoints fixed on a 2-dimensional Dbrane or D2-brane:

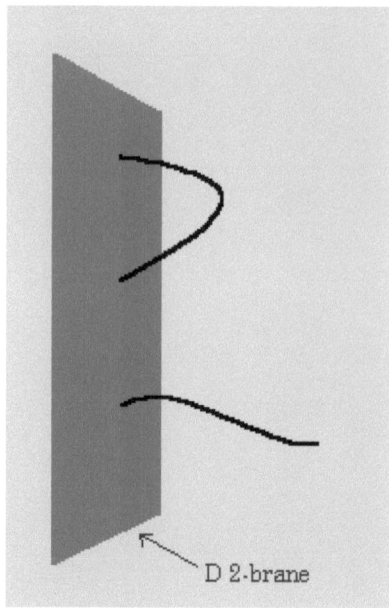

D 2-brane

D-branes can have dimensions ranging from -1 to the number of spatial dimensions in our spacetime. For example superstrings live in a 10-dimensional spacetime which has 9 spatial dimensions and one time dimension. Therefore the D9-brane is the upper limit in superstring theory. Notice that in this case the endpoints are fixed on a manifold that fills all of space so it is really free to move anywhere and this is just a Neumann boundary condition! The case p= -1 is when all the space and time coordinates are fixed, this is called an *instanton* or *D-instanton*. When p=0 all the spatial coordinates are fixed so the endpoint must live at a single point in space, therefore the D0-brane is also called a *Dparticle*.

Likewise the D1-brane is also called a *D-string*. Incidently the suffix 'brane' is borrowed from the word 'membrane' which is reserved for 2-dimensional manifolds or 2-Branes.

D-branes are actually dynamical objects which have fluctuations and can move around. For example they interact with gravity. In the diagram below we see one way in which an closed string (graviton) can interact with a D2-brane. Notice how the closed string becomes an open string with endpoints on the D-brane at the intermediate point in the interaction.

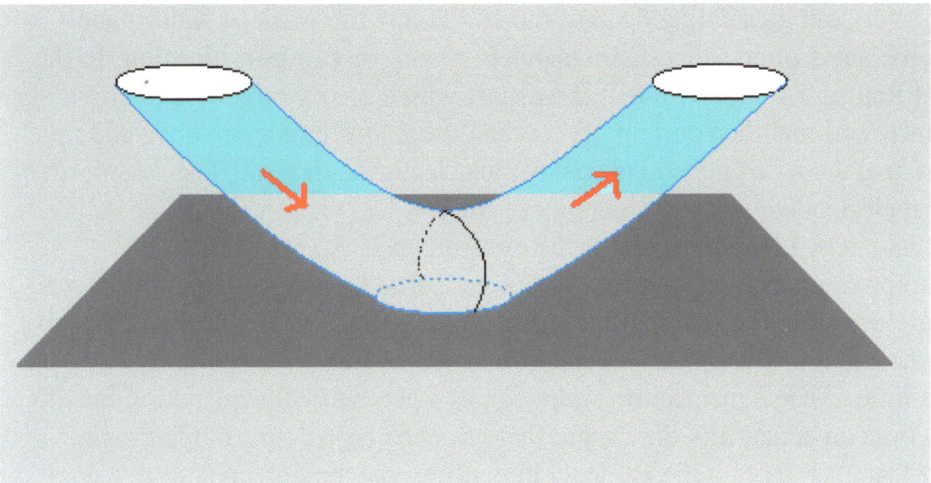

We now see that string theory is more than just a theory of strings!

There are two types of particles in nature - fermions and bosons. A fundamental theory of nature must contain both of these types. When we include fermions in the worldsheet theory of the string, we automatically get a new type of symmetry called supersymmetry which relates bosons and fermions. Fermions and bosons are grouped together into supermultiplets which are related under the symmetry. This is the reason for the "super" in "**superstrings**".

Superstrings live in a 10-dimensional spacetime, but we observe a 4-dimensional spacetime. Somehow we need to link the two if superstrings are to describe our universe. To do this we curl up the extra 6 dimensions into a small compact space. If the size of the compact space is of order the string scale (10^{-33} cm) we wouldn't be able to detect the presence of these extra dimensions directly - they're just too small. The end result is that we get back to our familiar (3+1)-dimensional world, but there is a tiny "ball" of 6-dimensional space associated with every point in our 4-dimensional universe. This is shown in an extremely schematic way in the following illustration:

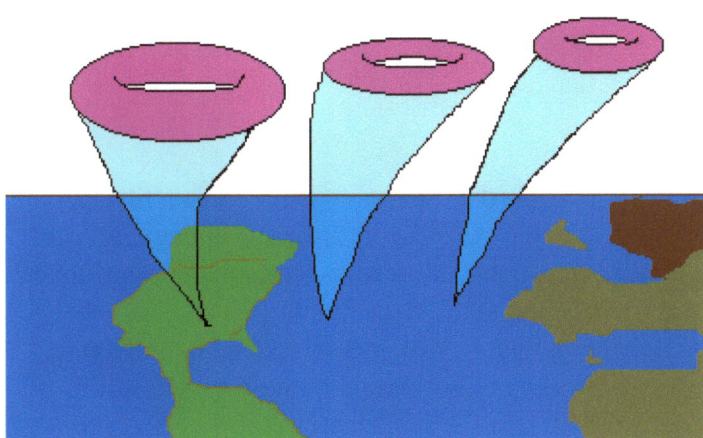

This is actually a very old idea dating back to the 1920's and the work of Kaluza and Klein. This mechanism is often called **Kaluza-Klein** theory or **compactification**. In the original work of Kaluza it was shown that if we start with a theory of general relativity in 5-spacetime dimensions and then curl up one of the dimensions into a circle we end up with a 4-dimensional theory of general relativity plus electromagnetism! The reason why this works is that electromagnetism a **U(1) gauge theory**. U(1) is just the group of rotations around a circle. If we assume that the electron has a degree of freedom corresponding to point on a circle, and that this point is free to vary on the circle as we move around in space-time, we find that the theory must contain the photon and that the electron obeys the equations of motion of electromagnetism (namely Maxwell's equations). The Kaluza-Klein mechanism simply gives a geometrical explanation for this circle: it comes from an actual fifth dimension that has been curled up. In this simple example we see that even though the compact dimensions maybe too small to detect directly, they still can have profound physical implications.

How would we ever really know if there were extra dimensions and how could we detect them if we had particle accelerators with high enough energies? From quantum mechanics we know that if a spatial dimension is periodic the momentum in that dimension is quantized, $p = n / R$ (n=0,1,2,3,....), whereas if a spatial dimension is unconstrained the momentum can take on a continuum of values. As the radius of the compact dimension decreases (the circle becomes very small) then the gap between the allowed momentum values becomes very wide. Thus we have a Kaluza Klein tower of momentum states.

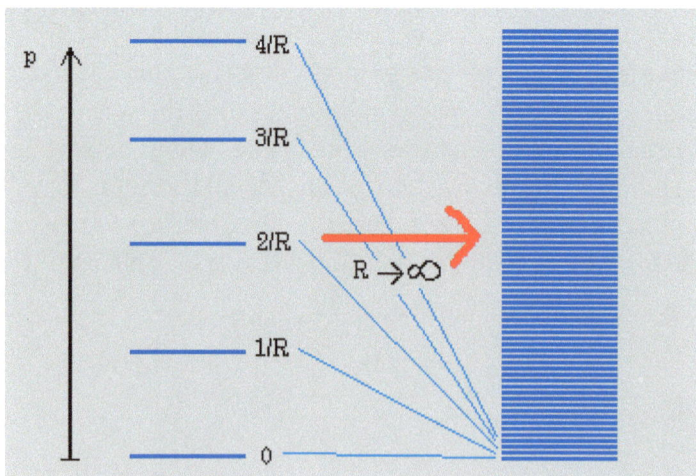

If we take the radius of the circle to be very large (the dimension is de-compactifying) then the allowed values of the momentum become very closely spaced and begin to form a continuum. These Kaluza-Klein momentum states will show up in the mass spectrum of the uncompactifed world. In particular, a massless state in the higher dimensional theory will show up in the lower dimensional theory as a tower of equally spaced massive states just as in the picture shown above. A particle accelerator would then observe a set of particles with masses equally spaced from each other. Unfortunately, we'd need a very high energy accelerator to see even the lightest massive particle.

Strings have a fascinating extra property when compactified: they can wind around a compact dimension which leads to **winding modes** in the mass spectrum. A closed string can wind around a periodic dimension an integral number of times. Similar to the Kaluza-Klein case they contribute a momentum which goes as $p = w R$ ($w=0,1,2,...$). The crucial difference here is that this goes the other way with respect to the radius of the compact dimension, R.

modes are becoming very light!

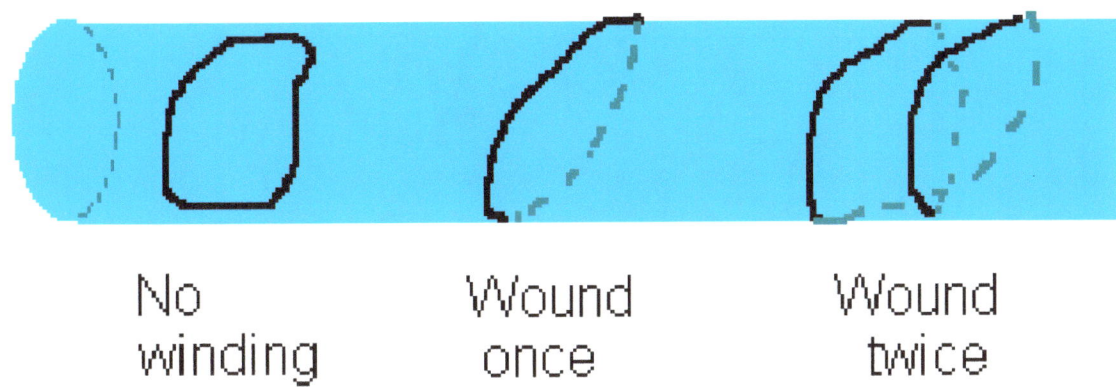

No winding   Wound once   Wound twice

Compact direction

Now to make contact with our 4-dimensional world we need to compactify the 10-dimensional superstring theory on a 6-dimensional compact manifold. Needless to say, the Kaluza Klein picture described above becomes a bit more complicated. One way could simply be to put the extra 6 dimensions on 6 circles, which is just a 6-dimensional Torus. As it turns out this would preserve too much supersymmetry. It is believed that some supersymmetry exists in our 4-dimensional world at an energy scale above 1 TeV (this is the focus of much of the current and future research at the highest energy accelerators around the word!). To preserve the minimal amount of supersymmetry, N=1 in 4 dimensions, we need to compactify on a special kind of 6-manifold called a **Calabi-Yau manifold**.

The properties of the Calabi-Yau manifold can have important implications for low energy physics such as the types of particles observed, their masses and quantum numbers, and the number of generations.

T-duality. This duality relates a theory which is compactified on a circle with radius R, to another theory compactified on a circle with radius 1/R. Therefore when one theory has a dimension curled up into a small circle, the other theory has a dimension which is on a very large circle (it is barely compactified at all) but they both describe the same physics! The Type IIA and Type IIB superstring theories are related by T-duality and the SO(32) Heterotic and E8 x E8 Heterotic theories are also related by T-duality.

S-duality. Simply put, this duality relates the strong coupling limit of one theory to the weak coupling limit of another theory. (Note that the weak coupling descriptions of both theories can be quite different though.) For example the SO(32) Heterotic string and the Type I string theories are S-dual in 10 dimensions. These means that the strong coupling limit of the SO(32) Heterotic string is the weakly coupled Type I string and visa versa. One way to find evidence for a duality between strong and weak coupling is to compare the spectrum of light states in each picture and see if they agree. For example the Type I string theory has a D-string state that is heavy at weak coupling, but light at strong coupling. This D-string carries the same light fields as the worldsheet of the SO(32) Heterotic string, so when the Type I theory is very strongly coupled this D-string is becomes very light and we see the weakly coupled Heterotic string description emerging. The other S-duality in 10 dimensions is the self duality of the IIB string: the strong coupling limit of the IIB string is another weakly coupled IIB string theory. The IIB theory also has a D-string (with more supersymmetry than the Type I D-string and hence different physics) which becomes a light state at strong coupling, but this D-string looks like another fundamental Type IIB string.

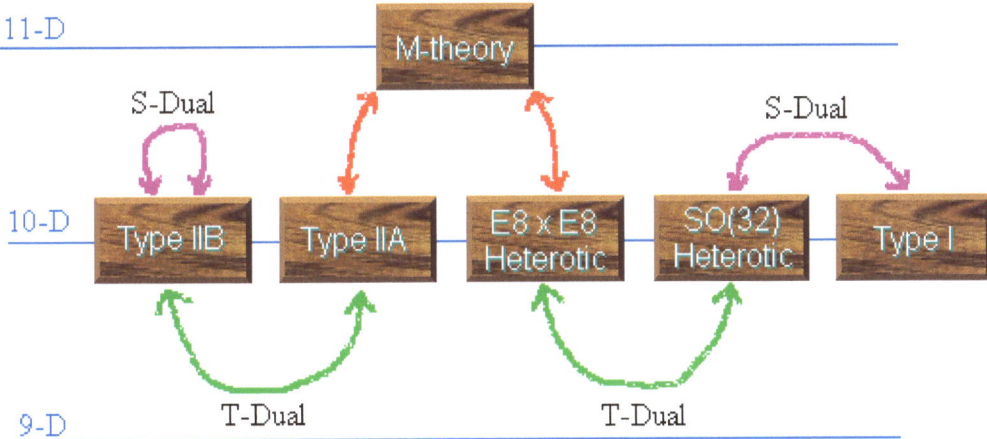

The dualities between the various string theories provide strong evidence that they are simply different descriptions of the same underlying theory. Each description has its own regime of validity, and in certain limits another description takes over just when the original one is breaks down.

M-theory is described at low energies by an effective theory called **11-dimensional supergravity**. This theory has membrane and 5-branes as solitons, but no strings. We can compactify the 11-dimensional M-theory on a small circle to get a 10-dimensional theory. If we take a membrane with the topology of a torus and wrap one of its dimensions on this circle this will become a closed string! In the limit where the circle becomes very small we recover the Type IIA superstring.

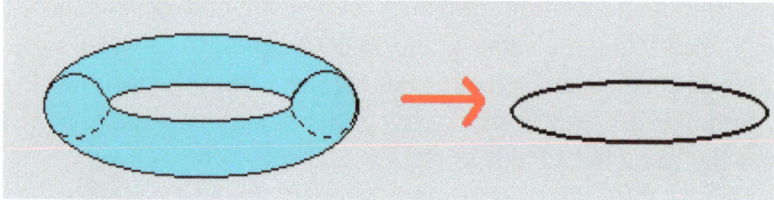

How do we know that M-theory on a circle gives the IIA superstring, and not the IIB or Heterotic superstrings? The answer to this question comes from a careful analysis of the massless fields that we get upon compactification of 11-dimensional supergravity on a circle. Another easy check is that we can find an M-theory origin for the D-brane states unique to the IIA theory. Recall that the IIA theory contains D0,D2,D4,D6,D8-branes as well as the NS fivebrane. The following table summarizes the situation:

M-theory on circle IIA in 10 dimensions
Wrap membrane on circle IIA superstring
Shrink membrane to zero size D0-brane
Unwrapped membrane D2-brane
Wrap fivebrane on circle D4-brane
Unwrapped fivebrane NS firebrand

The two that have been left out are the D6 and D8-branes. The D6-brane can be interpreted as a "Kaluza Klein Monopole" which is a special kind of solution to 11-dimensional supergravity when it's compactified on a circle. The D8-brane doesn't really have clear interpretation in terms of M-theory at this point in time; that's a topic for current research.

We can also get a consistent 10-dimensional theory if we compactify M-theory on a small line segment. That is, take one dimension (the 11-th dimension) to have a finite length. The endpoints of the line segment define boundaries with 9 spatial dimensions. An open membrane can end on these boundaries. Since the intersection of the membrane and a boundary is a string, we see that the 9+1 dimensional worldvolume of the each boundary can contain strings which come from the ends of membranes. As it turns out, in order for anomalies to cancel in the supergravity theory, we also need each boundary to carry an E8 gauge group. Therefore as we take the space between the boundaries to be very small we're left with a 10-dimensional theory with strings and an E8 x E8 gauge group. This is the E8 x E8 heterotic string. One I have tended to favor in one form or another.

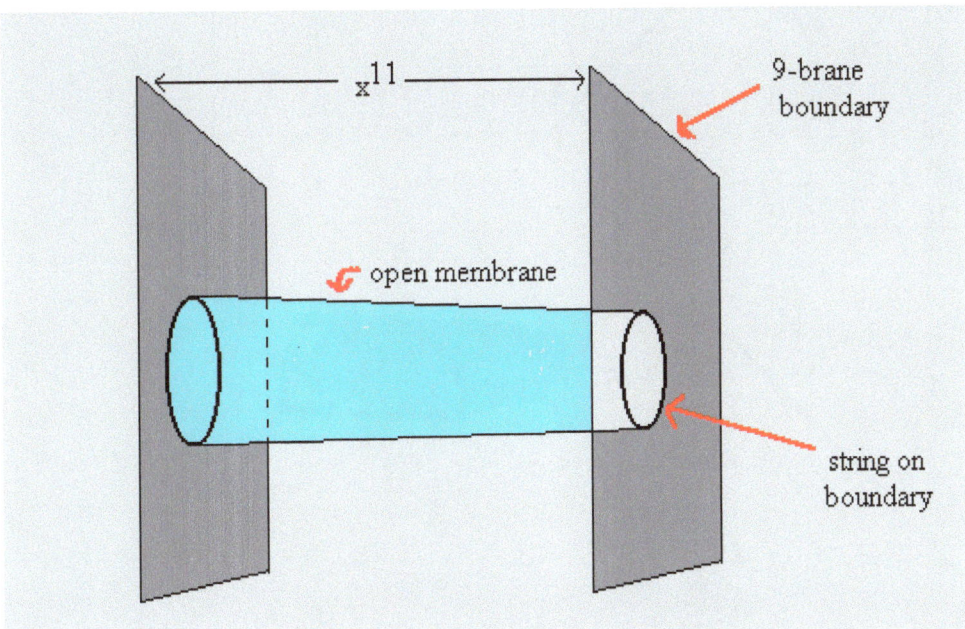

So given this new phase 11-dimensional phase of string theory, and the various dualities between string theories, we're led to the very exciting prospect that there is only a single fundamental underlying theory -- **M-theory**. The five superstring theories and 11-D Supergravity can be thought of as classical limits. Previously, we've tried to deduce their quantum theories by expanding around these classical limits using perturbation theory. Perturbation has its limits, so by studying non-perturbative aspects of these theories using dualities, supersymmetry, etc. we've come to the conclusion that there only seems to be one unique quantum theory behind it all. This uniqueness is very appealing, and much of The work in this field is toward formulating a solid fu;; quantum M-Theory.

The classical description of gravity known as General Relativity, contains solutions which are called "black holes". There are many different kinds of black hole solutions but they share some common characteristics. The event horizon is a surface in space-time which, loosely speaking, divides the inside of the black hole from the outside. The gravitational attraction of a black hole is so strong that any object that crosses the event horizon, including light, can never escape out of the black hole. Classical black holes are therefore relatively featureless, but they can be described by a set of observable parameters such as mass, charge, and angular momentum.

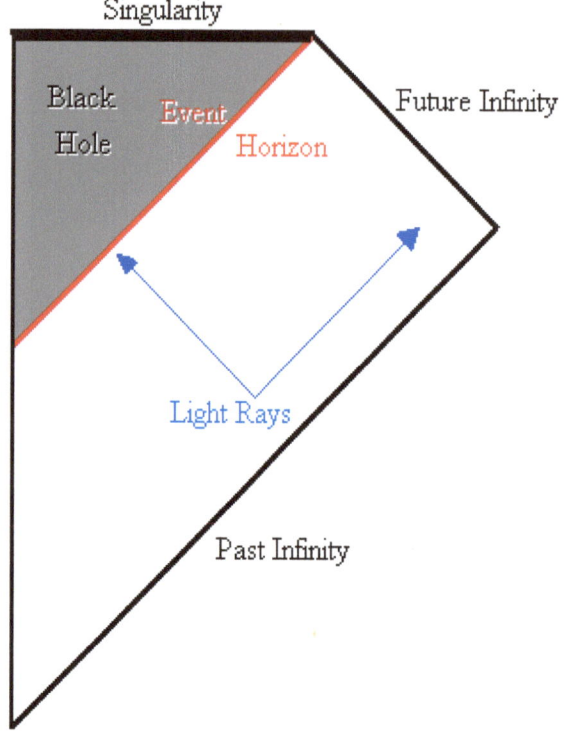

(explanation of this Penrose diagram)

Black holes turn out to be important "laboratories" in which to test string theory, because the effects of quantum gravity turn out to be important even for large macroscopic holes. Black holes aren't really "black" since they radiate! Using semi-classical reasoning, Stephen Hawking showed black holes emit a thermal spectrum of radiation at their event horizon. Since string theory is, among other things, a theory of quantum gravity, it should be able to describe black holes in a consistent way. In fact there are black hole solutions which satisfy the string equations of motion. These equations of motion resemble the equations of general relativity with some extra matter fields coming from string theory. Superstring theories also have some special black hole solutions which are themselves supersymmetric, in that they preserve some supersymmetry.

One of the most dramatic recent results in string theory is the derivation of the **Bekenstein-Hawking entropy** formula for black holes obtained by counting the microscopic string states which form a black hole. Bekenstein noted that black holes obey an "area law", dM = K dA, where 'A' is the area of the event horizon and 'K' is a constant of proportionality. Since the total mass of black hole, 'M' is just the rest energy of the black hole, Bekenstein realized that this is similar to the thermodynamic law for entropy, dE = T dS. Hawking later performed a semiclassical calculation to show that the temperature of a black hole is given by T = 4 k [where k is a constant called the "surface gravity"]. Therefore the entropy of a black hole should be written as **S = A/4**. Following recent pioneering work by Strominger and Vafa, it has been found that this entropy formula can be derived microscopically (including the factor of 1/4) by counting the degeneracy of quantum states of configurations of strings and D-branes which correspond to certain supersymmetric black holes in string theory. That is to say, D-branes provide a short distance weak coupling description of certain black holes! For example, the class of black holes studied by Strominger and Vafa are described by 5-branes, 1-branes and open strings traveling down the 1-brane all wrapped on a 5-dimensional torus, which gives an effective one dimensional object -- a black hole.

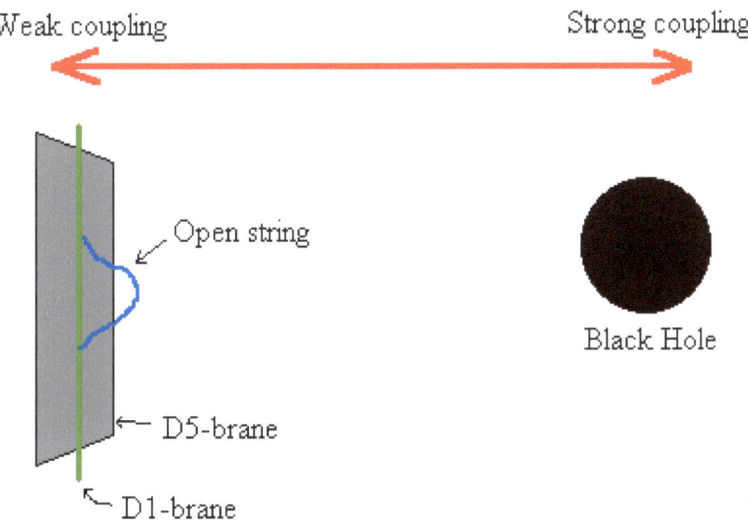

Hawking radiation can also be understood in terms of the same configuration, but with open strings traveling in both directions. The open strings interact, and radiation is emitted in the form of closed strings. The system decays into the configuration shown above.

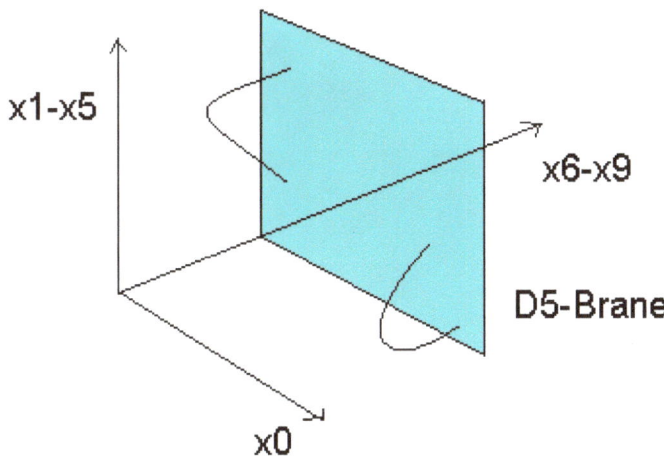

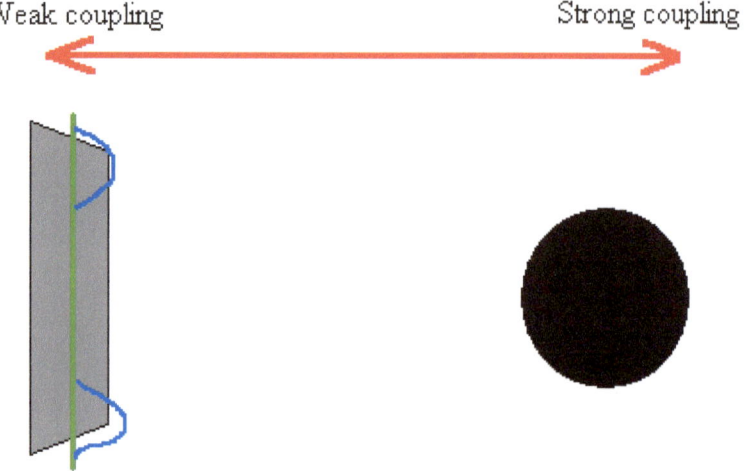

Weak coupling   Strong coupling

Explicit calculations show that for certain types of supersymmetric black holes, the string theory answer agrees with the semi-classical supergravity answer including nontrivial frequency dependent corrections called *greybody factors*.

D-branes" can be described simply as boundary conditions in perturbative string theory. A D-brane is the place where superstrings can end.

The world-volume of a D5-brane is a five dimensional hyper-surface in the theory which propagates in time x0, thus defining a six dimensional space-time.

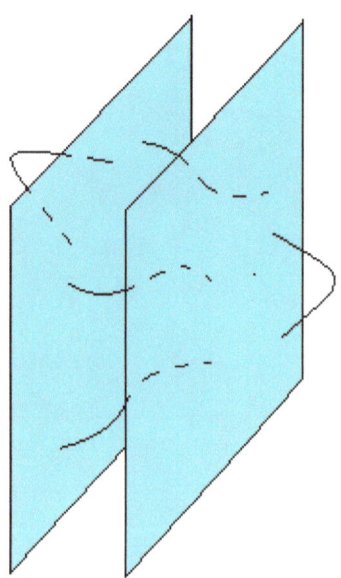

Multiple Branes and Gauge Theories

Here (left) we have a pair of D-branes, with superstrings stretching between them.

When the brane-brane strings are stretched, their vinrations have masses proportional to their length times their tension, and hence they do not generate addtional fields. If however the branes become coincident - that is, if they occupy the same place or position, new massless fields enter the theory.

These new fields coming from brane-brane string modes with coordinates parallel to the spacetime world-volume produce extra vectors in the theory. For N coincident branes ("N" means a certain numner, or any number), the gauge symmetry gets enhanced to U(N). ("U" being the symmetry group. The larger N is the more complex the group). There are also N scalars coming from the modes with coordinates transverse to the world-volume. These are the positions of the N branes in the transverse space.

SUPERSYMMETRY TOPOLOGY

Maxwell: Charge E moving along worldline γ, couples to the potential A:

$\mathrm{Exp}[i\int_\gamma eA]$

In M-Theory: Electrically charged membranes who's three dimensional worldvolumes couple to the C-field via

$\sim\exp[2\Pi I \int_\Sigma C]$

where the coupling C to all possible worldvolumes Σ summarises the gauge invariant information in C.

Shifted differential characters

More accurately, the coupling is

$$\sqrt{\operatorname{Det}\slashed{D}_{S(\mathcal{N})}} \exp\left(2\pi i \int_{\Sigma} C\right)$$

$$\Rightarrow \qquad [C_1 - C_2] \in \check{H}^4(Y)$$

Because of fermion determinants $[C]$ is actually a "shifted differental character" with quantization of its fieldstrength:

$$[G]_{DR} = a_{\mathbf{R}} - \frac{1}{2}\lambda_{\mathbf{R}}$$

$a \in H^4(Y, \mathbb{Z})$, λ = class of the spin bundle of Y.

Conclusion: "What is a C-field?" Partial answer:

$$\boxed{[C] \in \check{H}^4_{\frac{1}{2}\lambda}(Y)}$$

Relation to Differential cocycles

So the true gauge group is:

$$\mathcal{G} = \Omega^1(\mathrm{ad}P) \ltimes \check{H}^3(X)$$

with action on $\mathbf{E}_P(Y)$ is:

$$(\alpha, \check{\chi}) \cdot (A, c) = (A + \alpha, c - CS(A, A + \alpha) + \omega(\check{\chi}))$$

This is also <u>mathematically natural</u>:

Since we have a $\mathcal{G}$-action on a space $\mathbf{E}_P(Y)$ we can form the associated groupoid, which we regard as a category

$$\mathrm{Objects}(\mathbf{E}_P(Y)//\mathcal{G}) = \mathbf{E}_P(Y)$$

<u>The objects $\check{C}$ have automorphism group $H^2(Y, U(1))$.</u>

Hopkins & Singer's "differential cochains" are gauge potentials for abelian p-form gauge fields in all dimensions and degrees. They can be applied to the C-field of M-theory. The space of (shifted) differential cocycles $\check{Z}^4$ is also a category.

Theorem There is an equivalence of the two categories.

Proof: Both are groupoids with the same isomorphism class of objects, and each object has automorphism group $H^2(Y, U(1))$.

C-field action on manifold with boundary

In the case $\partial Y = X$ is nonempty the same formula applies

$$\Phi(\check{C}, Y) = \exp\left[i\pi\xi(\slashed{D}_A) + \frac{i\pi}{2}\xi(\slashed{D}_{RS}) + 2\pi i I_{\text{local}}\right]$$

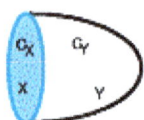

However, now there is a conceptually important distinction: the factor $\exp[i\pi\xi(\slashed{D}_A)]$ is a section of a $U(1)$ bundle with connection.

As is well-known from Chern-Simons theory, Φ is valued in a principal $U(1)$ bundle

$$\mathcal{Q} \to \mathbf{E}_P(X)$$

Each extension $\check{C}_Y$ of $\check{C}_X$ defines an element:

$$\Phi(\check{C}_Y, Y) \in \mathcal{Q}_{\check{C}_X}$$

and two such extensions satisfy the "gluing law"

$$\frac{\Phi(\check{C}_Y, Y)}{\Phi(\check{C}_{Y^0}, Y')} = \Phi(\check{C}_Y - \check{C}_{Y^0}, Y \cup \bar{Y}')$$

This property, in fact, defines $\mathcal{Q}$.

$\Theta_X(\check{C})$ is C-field electric charge

To interpret Θ let us "insert" an M2 brane wrapping

$$\sigma \in Z_2(X).$$

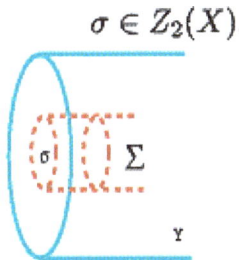

Consider the gauge invariance of $\Psi_\sigma(\check{C})$.

Flat gauge transformations $\check{\chi}$ with $\check{\chi} \in H^2(X, \mathbb{R}/\mathbb{Z})$ act on the wavefunction as

$$\check{\chi}(\sigma) = \exp\big(2\pi i \langle PD[\sigma], \check{\chi} \rangle\big)$$

Therefore, we interpret Θ_X as the C-field electric charge induced by the background metric and C-field.

Moreover, in the presence of membranes with spatial cycle $\sigma \in Z_2(X)$, the Gauss law is:

$$\Theta_X(\check{C}) + PD([\sigma]) = 0$$

Application 4: Spatial boundaries

Suppose now we have a spatial boundary $\iota : X \hookrightarrow Y$.

Impose boundary conditions on (A, c) via: $\iota^*(c) = 0$.

This boundary condition breaks the topological gauge symmetry $\mathcal{G}$ but is preserved by the usual gauge symmetry $\mathrm{Aut}(P)$.

Therefore, $\int_X \mathrm{tr} F \wedge *F$ is gauge invariant and therefore the E_8 gauge fields are indeed dynamical on the boundary!

• Note that this is spontaneous symmetry breaking of groupoids, not of groups.

In this way we can incorporate both Horava-Witten and Fabinger-Horava models.

The cancellation of E_8 gauge anomalies is manifest, and local.

Gravitational anomalies are more subtle.

We then encounter the idea that our worldvolume of 4 dimensions that we live in may be nothing more than a holographic projection of the information on the Brane itself. Outside of this brane and its Hologram in theory there would exist a different kind of space-time commonly refered to as Hyperspace. The normal assumption and format of M-Theory is that the extra dimensions are compact, the 4 dimensional worldvolume is built up from the different branes and the interaction of Strings and that Hyperspace is nothing more than some external space, usually considered anti-de Sitter that all this arose in. The interesting aspect of Fernando's own look at this is that, while some compacted, say 6 dimensional frame may exist, that hyperspace is not the odd exotic space, but actually just a very large version of our universe in which, from the perspective of our worldvolume everything is tachyonic. There would be differences in the expression of charge over there, there would possible be supersymmetric partners of regular matter over there, the vacuum state may be different: But it's just a huge superinflated counterpart of our universe when you boil it all down.

Going back to that D1-D5 Brane model I brought up: Our universe would be expanded from and wound one direction at right angle to the D1-D5 Brane, the other Universe would be expanded from and wound the other direction at right angle. The Other Universe superinflated, engulfed our brane and the original D1-D5 source. The D1-D5 source forms that compacted region, our worldvolume is basically an extended D3 Brane, and the Other universe is a superinflated D3 brane extension, add in time and you get 10 dimensions for either worldvolume and the compacted one.

MUSINGS ON THE EVOLUTION OF A COSMOS

A particular kind of string theory in ten dimension (called type II B string theory) can be compactified with five of the dimensions wrapped up as 5-sphere (S5) and the other five dimensions taken to describe a 5-dimensional anti de Sitter Space-time with negative cosmological constant (AdS5). The whole manifold will then be S5 × AdS5 with the metric on the AdS sector given by

$$ds^2 = dr^2 + e^{2r}(\eta_{\mu\nu}dx^\mu dx^\nu) \qquad \mu, \nu = 1, 2, 3, 4.$$

This string theory has an exact equivalence with the 4-dimensional N = 4 supersymmetric Yang-Mills theory. It is known that the latter theory is conformally invariant; the large symmetry group of the AdS5 matches precisely with the invariance group of Yang-Mills theory. The limit r →χ is considered to be the boundary of AdS space on which the dual field theory is defined.

If gravity behaves as a local field theory, then the entropy in a compact region of volume R3 will scale as S X R3 while indications from the physics of the horizons is that it should scale as S X R2.

The AdS space-time has a negative cosmological constant while the standard de Sitter space-time has a positive cosmological constant. Generally, there are few solutions to string theory which contains de Sitter space-time or even any solution to standard Einstein's equation with a positive cosmological constant.

The existence of a state of lowest energy or ground state is crucial for the classical stability of a physical system. In gravity, positive mass theorems establish the existence of a unique, zero mass ground state in both the asymptotically flat and the asymptotically anti-de Sitter(AdS) classes of Space-times.

The RS Model in general stipulates our vacuum state as a degenerate ground state of an original AdS space-time trapped by certain mechanisms like the Israel condition. But little discussion exists on how our brane with its positive energy and trapping mechanism of the Israel condition came into being in the first place.

We could predict that scattering of particles across an AdS spacetime will at the Limit

R→infinity

approach a flat space-time approximation[1]. So in general the outer boundary of an inflating AdS space-time will then approach more closer to a flat de Sitter

Space-time. So is there anything we could learn about the evolution of this outer boundary?

In a thin wall approximation we can represent the boundary in 5d gravity by a delta function source with some coefficient f(Φ), where the Φ is a bulk scalar field, the dilation parametrizing the tension of the wall or brane. While it is true that quantum fluctuations of the fields should correct this function or modify it, we might well ask that since this relationship exists is it possible the fluctuations of fields originally not trapped within the domain wall, but existent within AdS Space-time actually brought about the creation of the particles and fields trapped on that domain wall? The reason for asking this question will become apparent shortly.

As long as we consider quantum corrections only which modify f(Φ) while maintaining the bulk 5d gravity action we find that none of the corrections or actually shifting of brane tensions will destabilize the flat spacetime condition[2]. In fact they generate a tree–level distribution of energy ranges not unlike the one we encounter under the Standard Model with particle masses. Going a bit further we discover there is little reason for fine tuning since the restraint of a finite 4d Plank scale restricts the sign of f and the value of F/f on the wall itself which forces the value of the Cosmological constant to be in the range of an order of 1.

From this we could easily derive a picture or model of our universe existing as a brane or domain wall on the boundary of AdS space-time approaching a condition where R→∞ holds. It would follow then that our positive energy space-time with all its Israel condition trapped fields and particles is simply the natural result of quantum fluctuations of fields within the AdS space-time itself. If we should find that the trapped vacuum state has evolved with time then that evolution itself would be the result of quantum evolution of the AdS space-time fields.

Looking backwards in time toward the original evolution of the AdS space-time from the trapped perspective of the brane there would be little difference found from an observational point than that encountered in the normal Big Bang Model. We'd see a consistently flat space-time evolving from what appears to us as a singular point getting larger and larger until regions become casually disconnected. We'd have certain aspects that seemed to defy observation of the source and perhaps hint at the presence of unseen particles and forces because of our being limited in observation to only those processes that transpire on the brane within the domain wall.

As such we would suspect that a certain unseen or exotic aspect of energy must exist and that we perhaps have matter that exists in our trapped universe that we cannot observer directly, but, only through certain actions its presence would seem to be generating. In short we'd end up with a universe that observationally appears much like our own.

THE QUESTION OF TIME

Feynman recognized that time as we think of it becomes imaginary in the time reference we commonly use, because this time is totally indistinguishable from directions in space. if the universe exists in an unseen way without beginning or end, at right angles to regular time, then that time is simply more elementary and even more real than ordinary clock time. Thus it seems the term imaginary applies more accurately to our time. If the universe exists in another time reference where conditions are permanent or static, suddenly it doesn't matter that we humans so convincingly observe a beginning and a possible future end to our ordinary clock time, since the other time reference applies regardless of our sense of where we are in time. The universe could be said to exist before our clock time began, and after time ends.

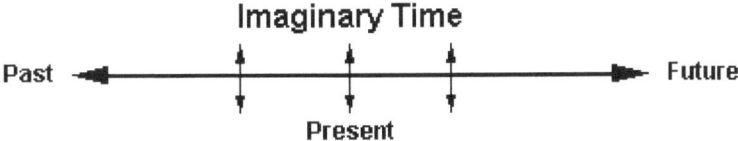

All moments share this time reference which has no beginning or end.

Imaginary Time

Past ◄————————————————————► Future

Present

This is exactly what one would expect from the before mentioned evolutionary model. Our time would be but a subset of a greater time that for all intents might extend into the past and future unlimited.

REFERENCES

[1] Joseph Polchinski, S-Matricies from AdS spacetime

[2] Shamit Kachru, Michael Schulz, and Eva Silverstein, Self-Tuning Flat Domain Walls in 5d Gravity and String Theory
http://www.slac.stanford.edu/pubs/slacpubs/8250/slac-pub-8337.pdf

On The Zero Point Field

The Equation Used

The stress-energy tensor on the right side of the equation holds all the information about the distribution of energy and mass in space-time, but Einstein tensor on the left, has only information about Ricci curvature which is only part of the picture. The field equation only tells us how the curvature at a point is affected by the matter and energy that are present at this point. It affects only Ricci curvature. Weyl curvature at a point is caused by matter and energy at other points.

$$R_{ijkl} = C_{ijkl} + E_{ijkl} + G_{ijkl}$$

is the Riemann curvature tensor in its full format. Thus,

$$E_{ijkl} = -\frac{1}{2}(g_{ik}S_{jl} + g_{jl}S_{ik} - g_{il}S_{jk} - g_{jk}S_{il}),$$

with

$$S_{ij}$$

is the traceless tensor

$$S_{ij} \equiv R_{ij} - \frac{1}{4}g_{ij}R,$$

and

$$G_{ijkl}$$

is defined by

$$G_{ijkl} \equiv -\frac{R}{12}(g_{ik}g_{jl} - g_{il}g_{jk})$$

and

$$R_{ij}$$

is defined by

$$R_{ij} \equiv R^k_{ijk} \text{ and } R \equiv g^{ij}R_{ij}$$

these all can be written as

$$R_{ijkl} = C_{ijkl} + \frac{1}{2}(g_{il}R_{jk} + g_{jk}R_{il} - g_{ik}R_{jl} - g_{jl}R_{ik}) - \frac{R}{6}(g_{il}g_{jk} - g_{ik}g_{jl})$$

We now have

$$E_a = F_{ab}\,u^b, \quad H_a = F^*_{ab}\,u^b$$

for the vacuum when observer is moving with time-like 4-velocity. This observer would feel the electric and magnetic components of the vacuum as

$$E_{ac} = C_{abcd}\,u^b\,u^d, \quad H_{ac} = {}^*C_{abcd}\,u^b\,u^d$$

Since this field is not purely electric or purely magnetic we have

$$C^{cd}_{ab} = 2u_{[a}E_{b]}{}^{[c}u^{d]} + \delta^{[c}_{[a}E^{d]}_{b]} - \eta_{abef}\,u^e\,H^{f[c}u^{d]} - \eta^{cdef}\,u_e\,H_{f[a}u_{b]}$$

which equals

$$C^{abcd} = (\eta^{acef}\,\eta^{bdpq} - g^{acef}\,g^{bdpq})\,u_e\,u_p\,E_{fq}$$

$$+(\eta^{acef}\,g^{bdpq} - g^{acef}\,\eta^{bdpq})\,u_e\,u_p\,H_{fq}$$

with

$$Q_b^a = \begin{bmatrix} 0 & 0 & 0 & 0 \\ 0 & \lambda_1 & 0 & 0 \\ 0 & 0 & \lambda_2 & 0 \\ 0 & 0 & 0 & \lambda_3 \end{bmatrix}$$

where

$$\lambda_3 = -(\lambda_1 + \lambda_2)$$

which in our case

$$Q$$

will contain real and imaginary components since the vacuum field is mixed and in constant flux as defined by quantum theory. As such it would be known as a radiative space-time. Since the vacuum as such has energy and energy is equal to mass it follows that the vacuum itself at some level tells space-time how to curve. But the components of both fields also constitute a drag effect for any object in motion. That drag effect has to be the origin of inertia and as such is the origin point of gravity.

If we take the space-matter tensor for a non-null EM field we have

$$P_{abcd} = C_{abcd} - g_{ac}F_{bk}F_d^k - g_{bd}F_{ap}F_c^p + g_{ad}F_{bt}F_c^t + g_{bc}F_{ax}F_d^x$$

$$+(\sigma + \frac{1}{2}F_{ij}F^{ij})(g_{ac}g_{bd} - g_{ad}g_{bc})$$

from which we get

$$P_{bcd}^h = C_{bcd}^h - \delta_c^h F_{bk}F_d^k - g_{bd}F_p^h F_c^p + \delta_d^h F_{bt}F_c^t + g_{bc}F_x^h F_d^x$$

$$+(\sigma + \frac{1}{2}F_{ij}F^{ij})(\delta_c^h g_{bd} - \delta_d^h g_{bc})$$

Thus,

$$T_{ab} = F_{ac}F_b^c$$

with

$$F_{ac} = s_a t_c - t_a s_c$$
$$s_a s^a = s_a t^a = 0, \ t_a t^a = 1$$
$$s_a s^a = s_a t^a = 0, \ t_a t^a = 1$$

where

$$x^i \rightarrow x^i + \xi^i dt$$

leaving

$$P^h_{bcd}$$

invariant which in turn implies motion of the medium and

$$\sigma = 0.$$

The motion of the medium is the expansion of the vacuum over time and as such it's motion is entangled with the CMB. An accelerated expansion implies an increasing motion to the medium. The differences in the field that should be evident would be those differences in the CMB itself. In such regions the vacuum's EM field would be moving at a different rate governed by the total energy present and it's effect on the excitation modes of the ZPF.

For accelerated expansion to occur there must be an increase of the amplitudes of the modes creating an increase in vacuum pressure countering gravity of the overall cosmos. A simple experiment involving two different size glasses of water on a vibrating platform will show that the larger glass shows higher amplitude waves. This in a crude format shows that accelerated expansion should have been expected as it is natural outcome of enlargement of the cosmic quantum sea.

In relation to particle travel or photons these higher amplitude waves amount to a longer travel distance generated at a quantum level which without quantum recalibration of our rulers results in C seeming to slowdown over time. But in reality it is our measurement of the distance of travel that is off and C has remained constant.

TOWARDS A SOLID QM THEORY

There exists a solution using some of De Broglie-Bohm Theory in conjuction with a modified particle picture that does account for certain quantum actions and non-locality if one accepts and off the brane fundamental sub quantum Dirac ether(not to be confused with a Newtonian Ether) of an absolute space and zero time frame De Broglie-Bohm Theory that in the end run will be shown to discard any absolute space-time frame in favor of a dual Lorentz Invariant frame.

This is currently the only interpretation in which all quantum effects are explained through causal continuous motions in space-time. Although it has not yet been generalized to relativistic domains it has been shown to fully contain the manifest (quasiclassical) world .

The Schrodinger equation can be treated the same way we would normally treat any other complex equation in physics just by separating real and imaginary parts and looking for the physical interpretation of each term. This was in essence the first hidden-variable theory proposed by de Broglie (as a direct consequence of his postulate of matter waves) and later developed by Bohm. In the 1950's when Bell was studying the hidden-variable theories he described the de Broglie-Bohm theory as having been `scandalously neglected'.

The basic idea is that the particle has a real position and momentum and that it is guided by the matter wave so as to reproduce exactly the normal results of QM. We can see how this comes about if we start with Schrodinger's equation for a mass m in a potential $V(r)$,

$$i^{h}/_{2\pi}\frac{d}{dt}\Psi = -\left(\frac{^{h}/_{2\pi}{}^2}{2m}\right)\nabla^2\Psi + V\Psi.$$

if we now define real functions of position R and S such that $\Psi = R\exp(iS/^{h}/_{2\pi})$, substitute and rearrange, breaking up real and imaginary parts we get,

$$\frac{dR^2}{dt} + \nabla\cdot\left(R^2\nabla\frac{S}{m}\right) = 0$$

and

$$\frac{dS}{dt} + \frac{1}{2m}(\nabla S)^2 + V - Q = 0.$$

for the two parts where

$$Q = -(^h/_{2p}{}^2/2m)\tilde{N}^2 R/R.$$

All we have done is to write Schrodinger's equation in terms of R and S but we can now see the physical ideas underlying the theory. Note that R_2 equals $|\Psi_2|$, which is the probability density of finding a particle over a certain distance. Also if we assume that the particle has a velocity $\mathbf{v} = S/m$ then we derive

$$\frac{dR^2}{dt} + \nabla \cdot \left(R^2 \nabla \frac{S}{m} \right) = 0$$

is just the continuity equation for the probability density and the second and third terms of equation are the kinetic and potential energies of the particle. From separating the variables in the time dependant case it follows that $[(d)/(dt)] S = -E$, the total energy, and

$$\frac{dS}{dt} + \frac{1}{2m}(\tilde{N}S)^2 + V - Q = 0.$$

is equivalent to the energy conservation equation with a modified potential. The last term, Q, is known as the `quantum potential' and the differences between quantum and classical predictions are due only to this additional term. Keeping this term small Classical emergence occurs and thus, becomes the whole model becomes classical.

THE SOLUTION

However, if we backup and consider the particle, not as some point set of

energy/mass being carried by the pilot wave, but as the whole pilot wave itself
we then get the same basic picture the above has been trying to generate but we
have abandoned the point particle picture of matter. The uncertainty principle
still applies to this interpretation but t no longer means that reality is unsure
about its self only that our observations of it are fundamentally limited. That limit
is imposed by our own macro lightcone state under this view. Those portions of
the pilot wave that are not C limited cannot be directly viewed in our classical space-time.
They imply a separate and external frame of reference with an
individual and separate version of C or time.

As to the issue of the double slit experiments, this wave, which is the particle,
would cross through both slits. The result is the same pattern we see duplicated
over and over again.

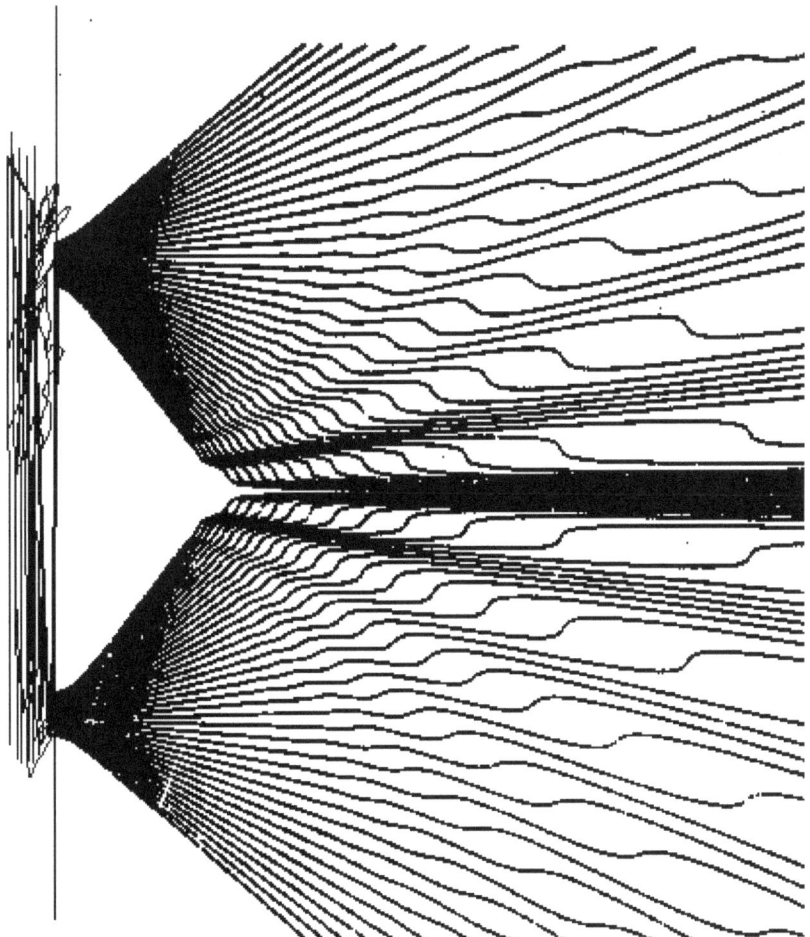

If we back up a bit, to my proposed zero point frame of reference(1) then the quantum potential has its basis in a sub-quantum Dirac ether of an absolute space and zero time frame. If we were sitting in the rest frame of a photon, watching it from above as it sits like a floating feather of EM energy in the sea of space-time. Then the two slits apparatus hurtles past at the speed of light. We'd get exactly the same pattern reproduced as above. The difference is there acceleration is caused by space-time curvature. Thus, the source of acceleration and the arrow of time is the curvature of space-time at a local level with its combined effect producing the over all limit on information flow. Also if it were acceleration being caused by curvature, then we'd get no emission of photons.

This then suggest how the underlying cause comes about from the fact that the quantum potential bends light and deflects the path of charged particles without making them radiate. But it also suggests that we have a hidden frame of reference in this universe with expanded light-cone states that only a modified version of SR/GR can take into account.

Quantum tunneling and non-locality effects, like those in entanglement, would be due to actual wormhole like connections through space, not time. These connections I believe are the long mentioned off the brane states from MTheory. All that is required from a QM perspective is to model the particle as the whole wave. From here, as shown above one naturally is led to such a conclusion. As one can see, this model fully abandons the point particle concept in favor of a more M-Theory like particle picture.

Now, if we look at some of Kant's ideas on space and time, as illustrated by the following diagram

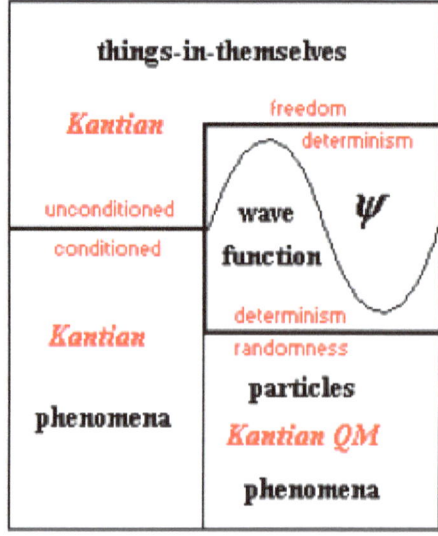

Kant's idea that space and time do not exist among things-in-themselves has been curiously affirmed by Relativity and quantum mechanics. In Relativity, time simply ceases to pass at the velocity of light: for photons that have travelled to us as part of the Cosmic Background Radiation, time has stood still for most of the history of the universe. On the other hand, quantum mechanics now posits **"non-locality,"** i.e. physical distances, and so the limitation of the velocity of light in Relativity, don't seem to exist, as demonstrated by Entanglement.

If, for instance, a positron and an election are both created from an energetic photon, the conservation of angular momentum requires that one be spinning one way, and the other the other. But the complementary spins are *equally probably* for each particle. Thus, in quantum mechanical terms, the wave functions of each particle separate without a discrete state being determined. The particles might then separate to even cosmological distances, but as soon as the spin of one particle is observed, the other particle *must* have the opposite spin, which means that the wave function has collapsed across those cosmological distances and *caused* the other particle to assume a predictable spin. If this occurs instantaneously, it would violate the limitation of the velocity of light in Special Relativity. This has now been shown to be true via experiments with entanglement(2). So at some level we have an already proven by experiment fact that SR is in need of a fundamental modification.

Now, contemporary physics states that no object should be able to travel faster than the speed of light. What they fail to state is that whether superluminal speeds are possible *in principle* depends on the **real** structure of the space-time continuum, which contemporary physics ignores, however. The above argument begins to address that issue of the real structure of the space-time continuum. Up until now we have had two basic choices:

Galilean Space-Time (GST)
Minkowski Space-Time (MST)

Briefly, whereas Galilean space-time allows the realization of faster-than-light speeds, at least in principle, Minkowski space-time does not.

It is important to note that without some definition of global time the physical *quantity* speed (and thus light-speed) has no definite meaning anyway. Why? Consider an example: Imagine an object moving from position A to B. Its speed v is given by the formula

$$v = \frac{\text{Distance}(A \text{ to } B)}{t(B, \text{finish}) - t(A, \text{start})}$$

Here, the start time $t(A,\text{start})$ and the finish time $t(B,\text{finish})$ are read off from two *spatially separated* clocks: one clock is located at point A and the other one at point B. Now, the difference of the two times in the denominator $t(B,\text{finish}) - t(A,\text{start})$ is an indefinite expression, unless there exists a rule how to synchronize both clocks, because clock B ignores the "current" time at clock A at first. But, in fact, the decision in favor of a particular synchronization rule is **pure convention**, because it seems impossible to send an "instantenous" (infinitely fast) message from A to B like "Initialize the clocks now!". Thus, the actual quantity of speed is conventional too, depending on the particular choice of the simultaneity definition.

Minkowski Space-Time does not know any absolute time which is physically meaningful. It was the revolutionary idea of Albert Einstein to give the notion of simultaneity a new definition. Especially, because all experimental tests to determine the motion with respect to some absolute space-time frame had failed, he decided to abandon the notion of absolute time at all. In the famous theory of relativity he postulated two principles which should hold for all physics:

1) All physical laws appear according to the same laws in all reference frames.

2) The speed of light is constant in all reference frames.

Second one is simply an assumption. It implies, in contrast to Galilean Space-Time, that simultaneity is not an absolute physical quality, but a relative one, depending on the motion of the observer (ie. the reference frame). However, it has to be emphasized that although the existence of a physical absolute time (or, equivalently, a preferred reference frame) could not be established by experiments, the theory of special relativity does not disprove it either.

So, let's go back to the above discussion of a decent QM model. Under Bohm's version one encounters aspects of the pilot wave that exceeds the local established value for C, and modern experimental aspects of entanglement we also encounter something similar. So let's make the assumption that statement 2, is simply a limited case finding and that there are aspects to space-time in which this limited finding does not apply. What effect would this have?

The general coordinate transformation from this particular reference frame R to a general one R' (primed coordinates) reads

$$x' = \gamma\,(x - vt), \quad y' = y, \quad z' = z$$
$$t' = \gamma^{-1}t + S(x')$$
$$\text{with } \gamma = \left(1 - \frac{v^2}{c^2}\right)^{-1/2}$$

where the relative speed v of R' with respect to R is chosen to be parallel to the x-axis. The transformation properly expresses the apparent contraction of moving rods (Lorentz- Fitzgerald contraction) and the slowing of moving clocks (time dilation). The function $S(x')$ simply determines the notion of simultaneity in frame R'. Generally, $S(x')$ can be an arbitrary function, but it is convenient to impose $S(0) = 0$ such that the clocks of the reference frames R and R' are synchronized at the origin $(x,t) = (0,0) = (x',t')$. Furthermore, in order to avoid acceleratory effects, one usually imposes that $S(x')$ is linear in x', ie. $S(x') = s\,x'$.

But, if we substitute a different value for C into the above equation we maintain the Lorentz Invariance, at the expense that C is no longer a constant. However, if we try to project this altered situation back into a frame were C has the value we measure it at globally, we loose some of the information in the transfer and encounter a situation where the energies of such particles within the altered frame system crossing into our regular region would seem to be higher than we can account for due to their measurable velocity. We would also loose some of the information in any quantum accounting which would give the appearance then of action at a distance, or non-locality. So, by simply modifying Relativity to include states who's value for C is higher than our own we have also provided a solid solution to the Non-locality problem proposed by Bell(3).

It can be shown that Einstein's second postulate is equivalent to setting $S(x') = -v/c^2\ x'$, so that one obtains the well known Lorentz transformation equations

$$x' = \gamma\,(x - vt), \quad y' = y, \quad z' = z$$
$$t' = \gamma\left(t - \frac{vx}{c^2}\right)$$

with the speed of light $c' = dr'/dt'(r=ct) = c$ constant in all frames. Thus, from the viewpoint of relativity, all reference frames are completely equivalent. And when we substitute in a new value for C we find this to still hold, until we try to compare systems who's local velocity for light, or information transfer is different.

The Traditional QM approach has suffered this problem when it comes to bringing gravity into the fold. For most of us there is an inevitable logic in this sort of structure: an event at **A** causes an effect on an object at **B**, which later causes a reaction in a separate object at **C**, which may or may not have an effect on **A** at some time in "the future". That **C** comes objectively after **A** and is determined by it has the force of an axiom (the causal order **A** - **B** after all *defines* what we mean by "after"), and that objects like **C** are "made of" more fundamental component objects and processes like **A** is not even a "good theory" - it is more so obvious, indeed, that one needs constantly reminded that the formal expression of these concepts in the doctrine of micro reductive determinism was the result of centuries of European philosophical thought.

This is a cultural invention because quantum theory (in principle if not quite yet in fact) quantizes **gravity** along with the other forces and by thus quantizing space-time (which is Einstein's gravity field) promises to deny us any way of making this matrix of space and time precise enough to define when and where an isolable event takes place, and the non-locality of quantum systems calls into question the very meaning of an "isolable event". The particular axiom which quantum gravity threatens to undermine, therefore, is the axiom that the **causal order** and the **space-time order** are the same thing.

But, this view I have proposed keeps the casual order of events in this universe stable, while explaining the whole issue of Non-locality in general. In this model of quantum theory, casual order is preserved as is the space-time order.

However, by the introduction of Bohm based views into a model where the wave is itself the particle, we have automatically introduced on the surface a dual time frame. The reason is that with Bohm's original pilot wave the particle was present within the wave, being carried along by it. Depending upon the particles mass its local velocity would always be less than the overall wave's velocity. Even with a zero rest mass particle like the photon, this would still be true. Thus, when we picture the whole wave as the particle, we automatically introduce one state with a C limited time and another with an expanded light-cone state.

Now, if we wish to include a proper string theory based view into this, we'd find these two states to be the result of the mixing of two separate time states themselves. So in essence we'd have to introduce what actually amounts to a Triple time format into our model. One Time state is the resultant, the extended one is the compacted regions time, and the third would be the full off the brane time state which is negative. However, for the case in point of introducing this basic model I have simplified it to the brane and compacted region only.

Reference

1.) Author Journal of Theoretics Vol.4-5 Non-Orientation of Space-Time Proves Mtheories Compacted or Embedded Regions.

2.) Dr. Serge Haroche, ENS & College de France (ITP 9-26-01)
http://online.itp.ucsb.edu/online/colloq/haroche1/pdf/haroche1.pdf

3.) J.S. Bell, *Physics* 1(1964), p, 195.

The Anthropic Principle and Intelligent Design

The anthropic principle arose due to Brandon Carter, who articulated the anthropic principle in reaction to the Copernican Principle, which states that humans do not occupy a privileged position in the Universe. Carter said: "Although our situation is not necessarily central, it is inevitably privileged to some extent." "See:B. Carter, IAU Symp. 63 (1974) 291." There has always been a clash between fundamentalists type Christian views with man having a central position and secular views that tend to deny this. In more recent times, given the inability of groups like Christian Science Research to argue around the age of the earth and the Cosmos, they have reverted to the idea that all of our more recent odd observational findings like accelerated expansion show the local galaxy and earth have only recently begun to exist a gravity well. This is again another attempt at turning back the clock of science and going back to us being the center of the universe.

Now the problem with this is that while even under the BB model the universe all started from one point or cosmic egg as it is sometimes called and as such even with expansion would have at least a general center of mass effect via gravity since the expansion is not totally uniform over time even if one knew the exact position at this moment of every body of mass that would at best give a center of mass for the present that is different from the original. There simply is no way to point and say this is the center of the universe and the earth is here in relation to that center and as such all the odd effects are from us exiting a common gravity well.

A couple of hidden "mistakes" or rather assumptions are also buried in their idea:

1.) Age of Universe is off because we are only now beginning to exist that gravity well.

2.) If Earth was at center of Universe then Bible must be correct.

Instead of actually following established scientific methods they start with an outside assumption of the Bible being correct and then try to turn the evidence science gathers around in their favor while ignoring anything from science usable as disproof of their theory.

There are three different kinds of anthropic principle: The strong version, very weak version and weak version.

The very strong version states that everything in our universe has something to do with humankind. But in general even though that idea is unsound it is the one I most often see the Christian community resort too. The very weak version takes the very existence of our humankind as a piece of experimental and observational data. For instance, in order for a person to exist the way we do with the products of radio-decay, the life-time of a proton must be at least 10_{16} years is seen as evidence that life seems to hold an important position. Now the weak version postulates that there are many regions in the universe. This usually in one form or another the multiverse concept. It can be stated that there are many universes or that our universe is simply part of a greater whole with all sorts of vacuum states and physics being possible. It just happens that in the region we are dwelling all physical laws, physical constants and cosmological parameters are such that clusters of galaxies, galaxies and our solar system can form, and humankind can appear. This is rather we just happen to exist kind of path.

This later approach suffers from an attempt to push the answer far enough off that the question becomes non-answerable while the first approach is based upon pre-assumptions founded in faith, not science. The middle ground idea which allows us to be part of the observation and experimental evidence is more scientific even if it at present does not provide us solid answers. The later also even itself demands a first cause even if that cause is natural processes and as such still involves an origin point and an explanation of why everything is the way it is. The first cannot be considered as science since it violates so many scientific methods with the pre-assumptions to begin with. The later itself fails by its own avoidance of the answer even if there is plenty of theory behind the central idea.

One math based argument against the strong position is: Let z be the red-shift when galaxies form, the matter density is this

$$\rho_m(z) = (1+z)^3 \rho_m^0 \sim 100\rho_m^0$$

This requires

$$\rho_\Lambda \leq 100\rho_m^0$$

For a time scale to occur. Since we already know that expansion of the cosmos is not everywhere uniform then certainly earth does not hold a central place in the cosmos even if life may on the surface seem to and the age of the universe is no less than that required to have the proper energy density and the start of the universe requires at least that type of over all energy density. We generally assume a primordial energy density variance of

$$\delta\rho/\rho \sim 10^{-5}$$

But if we let

$$\delta\rho/\rho$$

be a variable we usually find

$$\rho_\Lambda \leq X\rho_m^0 \text{ and } X \gg 100$$

Under which dark energy is not a result of life or us. Adding to this

$$\rho_\Lambda/\rho_m^0 = 9 \text{ then } T = 1.1H_0^{-1}$$

For the age of the universe in itself does not dictate a central place for us on the surface. Our time scale is but a brief instant of the whole time scale in a Universe of variables with our position in that universe not being a provable central one. Even presupposing a Creator which I have no argument with the evidence that exists does not simply show that Creator to have created everything for the purpose of us alone. So I simply must reject that form of the whole principle to begin with.

The central problem is Creationists want science to prove God exists. By definition God is external to his or her's creation. Science can only study nature or by definition what God created. It cannot nor does it have the means to study something outside of nature. Yes, one would suppose if there is a Creator that creator left his signature in that creation. The problem is that signature if it exists can be said at this point not to support the type of Creator most Creationists believe in.

In the same token, science can only use the evidence it has to make conclusions. Going beyond the evidence even if there is that lack of a signature requires assuming something beyond the evidence we can study. This is where the Atheist crowd wrongly assumes science supports their position and in a fashion takes the same path the Christians do. They started with a pre-assumption and try to force the evidence to support them.

I am what I am. A scientists at heart. I am at best by science limited at present to an agnostic position by science. However, as an individual I can think outside the box of science. I would note that at a quantum level our Universe in which we exist acts very much like a giant quantum computer. It also has aspects similar to a holographic system. The question that always comes to my mind is who programmed the computer? Given, in our limited knowledge all computers no matter the scale require an outside programmer that question seems a valid one. Yes, I could simply postulate it programmed itself through natural means. The problem is that is as bad as the multiverse solution. It avoids the answer via attributing an unknown process for self-programming. Even if the origin is natural one still is required to account for that natural process.

There have been a number of attempts over the years to account for the fine tuning required by the universe we live in. One is *Supersymmetry with broken symmetry being the cause or origin of the primary fine tuning. The second is Modifying gravity to somehow account for the fine tuning. The third major approach is that of the wave function of the Universe.*

The third solution lacks any precise answers. For one, if we do live in a multiverse, if there are interconnections between all the different universes, if gravity from one effects the other then the whole simplistic wave function itself requires modification. It would be the case of pre-built in geometry at play. The first instead of providing one solution provides too many solutions none of which have exactly matched our current vacuum state to even begin to account for such fine tuning. At best it only points towards a possible mechanism. Modifying gravity while a valid path is itself only part of the solution. In fact, the other two paths rather modify gravity to begin with.

The first actually gave rise to the multiverse approach. String theory, derived from it showed a multitude of vacuum states. The argument was that perhaps the whole universe is populated with all these states. In other words, one needs a mechanism to produce different universes in a multiverse. The mechanism supplied was first string theory and later extended into Brane theory where our universe is but one possible brane connected via hyperspace or the bulk to other possible branes. It was found that two branes via boundary conditions could provide fine tuning. However, a stability problem occurs in this simple picture requiring its own fine tuning.

Men like Steven Hawking pointed out that the wave function of the universe is one populating method. Different observers live in their different histories, and they are summed over in the no boundary path integral. In other words, observers in different universes live in different decohered branches of a single wave function.

Under this given certain conditions required for life we happen to exist in one branch capable of life. It has in challenge been pointed out that a universe with conditions all the same as our universe except the CMB temperature is higher is more likely, the probability of its occurrence is higher than for our type. So it leaves an unanswered question in itself concerning why with the odds higher for that type of space-time do we find ourselves in a de Sitter space which is a resonant state in the multiverse.

That resonant state seems itself to require more than one vacuum state to produce the resonance against. That being the case it could be proposed that the mechanism is simply that ours is the natural resonant state of a multiverse of possible vacuum states.

It is then as if all the quantum probability conspires to produce this type of universe in which life is possible. This is where I have often found the answer to Einstein's does God play with dice question to be that He does play with dice,. They are just a loaded set of dice with a predictable outcome.

One thing that can be said is space-time's life time can not be longer than the Poincare recurrence time if we view this space-time has finite dimension of the Hilbert space. That remains one solid ground from science we already know. So the question is valid on what existed before the beginning of those finite dimensions?

If we follow the Bayesian statistics approach we find

$$P(\text{theory } x|\text{selection}) = \frac{P(\text{selection}|\text{theory } x)P(\text{theory } x)}{\sum_y P(\text{selection}|\text{theory } y)P(\text{theory } x)}.$$

Basically suggesting it's own version of the original question.

The whole core of the problem is we as observers alter what we observe. If there was no observer would the outcome be the same? It all becomes rather like if a tree falls in the woods and no one is there to hear it does it make a sound sort of question.

The simplest solution is usually the correct one. That being the case I find it just as valid if not even more valid to assume there must have been a first cause. The idea of a first cause can be seen as either intelligent or mechanical depending upon your preference. Given the lack of Scientific grounds to prefer the type of Creator of classical Christianity one could simply assume that this first cause being outside of nature cannot be directly studied.

However, given the fact that our universe seems to include in its program a gearing towards life that in itself seems to require Intelligent design. As such, the idea that our Creator could be a being or group of beings of intelligence is not that far fetched. But what is far fetched is the assumption that all of this was set up so human's could evolve. The assumption that it was geared so life could evolve is acceptable. But, again the central position of man has to be rejected simply because of the fact that the geological preserved record shows us to be rather late arrivals as far as life goes.

The question then becomes could we find evidence in the biology of life itself that shows that program at work? At this point I have derived the whole idea of Intelligent Design without having to assume either a Creator or any aspects of that Creator. I have not derived this path by prior assumptions as the current type of Intelligent Design tends to follow. I can then on a much firmer scientific basis begin to look closer at the evidence from biology to see what can be gleaned.

As I pointed out before Steven Hawking postulated that the wave function of the universe is one populating method. Different observers live in their different histories, and they are summed over in the no boundary path integral. In other words, observers in different universes live in different decohered branches of a single wave function. Under this given certain conditions required for life we happen to exist in one branch capable of life.

It has in challenge been pointed out that a universe with conditions all the same as our universe except the CMB temperature is higher is more likely, the probability of its occurrence is higher than for our type. So it leaves an unanswered question in itself concerning why with the odds higher for that type of space-time do we find ourselves in a de Sitter space which is a resonant state in the multiverse. That resonant state seems itself to require more than one vacuum state to produce the resonance against. That being the case it could be proposed that the mechanism is simply that ours is the natural resonant state of a multiverse of possible vacuum states. It is then as if all the quantum probability conspires to produce this type of universe in which life is possible.

The pseudo-Prior Geometry the quantum brane model I employed tends to produce a faster growth rate for cosmic structures like Galaxies. Something that has been observed very recently. Since one first needs structures like Galaxies in which planets can form and eventually life this model would tend towards a quicker development of life since everything else basic to life develops at a faster pace. But, the model does not supply an exact answer for when life first started.

I think the answer to that question goes back to multiple variable quantum states. Those against evolution have always pointed out that the high odds for certain events to transpire tend towards a negative on the question of evolution. However, if again, life or the start of life was a result of a resonant state of wave functions then those odds diminish to almost nothing against life starting given the basic building blocks are present.

We know the post BB universe was composed of mostly lighter elements. But if galaxies formed quicker than a normal BB model would predict then supernova production of heavier elements would itself take place faster under such a modified BB model. It is these heavier elements that are required along with lighter elements to yield the building blocks of life and DNA/RNA. I would suspect then that the building blocks and life itself occurred vastly earlier on than most evolutionary models tend to suppose. I think these seed forms of life became scattered throughout the cosmos as it grew and evolved. They simply awaited the right conditions to spread in abundance and fill what ever opening became available.

This being the case, our own Earth's evolution has probably had an interesting history. I would speculate that even if life first arose under earth Mark 1, the earth prior to the formation of the Moon, that such life was wiped out during that formation process. It was only after Earth Mark 2 had reformed and cooled down that a place for life to start again arose.

That being the case I think one can propose that the seeds of life exist out in space itself.

One place of possible hiding would be the Ort cloud, which from time to time has had objects enter into the inner system, crash into planets, leave dust behind in their tails, etc. Another would be the rocky, heavier mass objects that have from time to time impacted planets as they developed and aged.

These seeds of life then would all seem to share a common genetic format meaning that life everywhere should be based upon a common set of DNA elements. This in turn implies a prior-built in common set of DNA for all life that does arise. These seeds of life or simply cell organisms simply find a suitable place to take root, fill all possible avenues via diversity along evolutionary paths, and further evolve, flourish, and multiply as conditions allow. It is also possibly that the original simple cell organisms come in multiple variety to begin with even though their basic DNA patterns have a common basis.

This being the case, even though the mechanism of life and the timeline involved is not the same as the standard Creationist model, it does have elements in common with that model. It does have an earlier starting point for life and for the structure of the Universe. It does imply a reasoning for all life as we know it seeming to have a common DNA pattern upon which everything is built. It maintains that life is the normal instead of an accident of chance. It also implies that wherever the conditions are right life will arise and evolve. It also shows that even though we can prove the building blocks of life can be generated in a lab that there are elements of the requirement for life that we simply do not have the ability to generate at the present in a lab.

This leaves us as Scientists being able to study the planetary evolution of life to determine the how and why certain changes occurred, but, not with the ability to fully study the very start of life itself at least until such time as we can fully duplicate the conditions and variable quantum states in a lab under which such life first started. I would suggest that future attempts at exploration to search for life might want to consider capture of comet and other solar material to search for the seeds of life itself.

The False Idea of an Absolute Time and an Absolute Now

Many people via common sense watching everyday events get the idea there is an absolute time and an absolute now. The Problem is the physical workings of this Cosmos tend to do away with such silly notions, especially at the Quantum level. Taking the Liouville Equation at face value I explore the dual time frame effect upon quantum events and show there is no absolute now or absolute time only a connection of the past and the future.

The Liouville Equation is:

$$\dot{\Lambda}_a = i[H_a, \Lambda_a] + \sum_\beta g_{\beta a}^* \Lambda_\beta g_{\beta a} - \frac{1}{2}\{\Lambda_a, \Lambda_a\},$$

$$\dot{\rho}_a = -i[H_a, \rho_a] + \sum_\beta g_{a\beta}\rho_\beta g_{a\beta}^* - \frac{1}{2}\{\Lambda_a, \rho_a\},$$

where

$$\Lambda_a = \sum_\beta g_{\beta a}^* g_{\beta a}.$$

Now,

$$g_{a\beta}$$

depends upon what we in Physics call York time. The Hamiltonian describes the exchange of forces, of the system, and the first describes the exchange of bits of information. If the

$$g_{aa}$$

Term vanishes then it is via the overlap of multiple time state-vectors. The equations require this to be so. The simplest origin of multiple time state-vectors is the one proposed by Wheller involving advanced and retarded waves or signals, which is what I have already introduced before. Since our vacuum has Lorentz invariance and a certain global value for C only one of these waves can originate in our vacuum. The other must stem from an external vacuum who's value for C is different from our own.

This implies our universe has multiple time state-vectors for every event. In short, the past and the future determine the now. Discounting quantum probability for a moment as far as outcomes go the overall system is rather like a loaded set of dice. However, on a larger scale when you examine both time state vector sets our present is governed by the future because that future has already worked itself out in that expanded C vacuum state. So in essence the past choice we made in this time we live in reflects itself backwards from the future to re-enforce the result we call the present. At each choice or step in the past even though there are many different prospects, once a choice is made the next step we call the present becomes pre-determined via the echo backwards from the future. But even this past point was itself predetermined by a further back past and predetermined itself via that echo backwards.

In short, the question has been asked about the Cat in the Box. The Cat is in the Box because a past event triggered it to end up in the box. The outcome for the Cat was itself predetermined by the past and the future. One only has to open the Box in the present to determine what that outcome was. But even if one choice not to open the box at the specific moment even though the cat could be alive or dead it is only in one of those states and if left in the box the Cat will end up dead anyway making the whole Cat in the Box argument useless.

Circular logic it seems is what makes the universe go round. But the circular logic also serves a higher purpose. Think of it as an error checksum process. This twin set of times prevents information decay in our Universe as a giant quantum computer. The information going into the system is backed up with an information flow from an already worked out future. The real fabric of time and space is in the end determined by a fabric woven by every action that ever has existed and ever will exist. That fabric then is woven by connections more than anything else. There is no absolute time or absolute now.

This then introduces us to our next subject of if Pre-cognition can exist in this Universe.

Precognition is psychic knowledge of something in advance of its occurrence. It comes under those subjects generally discussed in the field of parapsychology, where scientific methodology is applied to the subject of ESP in general. However, many have often raised the question of just how Scientific is such a subject when it includes action (one's thoughts) being influenced by another action or event at a distance. I believe part of this question can be answered via the field subject of Quantum Mechanics which we have already looked at. It goes back to the whole dual time frame and those advanced and retarded signals at a quantum level. Since our minds, functioning as they do as electro-mechanical computers, all at some fundamental level utilize quantum effects, it is possible for our minds to under certain conditions pick up on events of a future nature long before those events take place.

In the original theory the advanced and retarded waves are replaced with our normal C lightcone state and a much faster C lightcone state. In essence, you end up with the same effect as originally proposed without the need of any backwards in time wave. The result stems from a dual lightcone where one C is our normal 186300 Miles Per Second and the other state is a C where on the order of C^3 would be the new velocity of light. That compacted hidden region is a Universe where our whole cosmic history took place in an instant, while on our side of the universe everything moves and transpires at a much slower rate.

Notice I mentioned certain conditions to all this. At the current time any solid scientific research in this area has more originated out of the Para-Psychology branch of Psychology than any hard physical science type lab experiments. What I am proposing actually holds more in common with the older Wheeler-Feynman Absorber Theory as mentioned before and illustrated below.

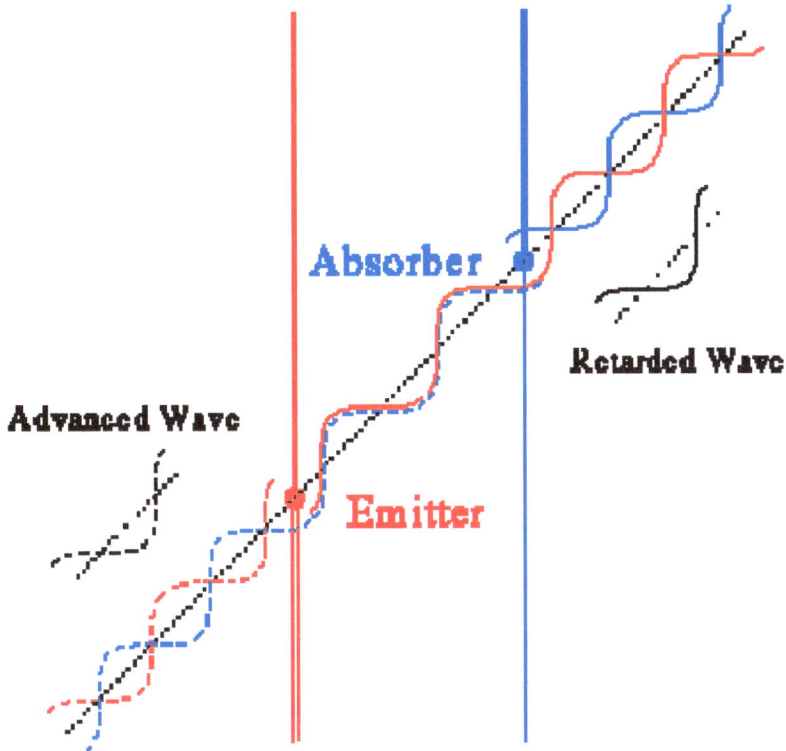

In essence, when you look at how our Cosmos may function at a quantum level the whole idea of Pre-cognition becomes vary soundly standing on a lot of different quantum theory taken as a whole. In my Book Incident at Chanute complete edition I mentioned the percentage of correct remote viewing recordings of both our Military and CIA and the Russian Program. In general, while again we have more of a psychology type research being provided as evidence such can exist, the evidence from experimental avenues seems to favor that some form of pre-cognition does exist at least in certain individuals.

A NEW MODEL

In the case when we have two world sheets, one with positive tension at y = 0 and a second one at y = d with sheets satisfies

$$\left(\frac{1}{a^2}\Box^{(4)}\bar{h}_{\mu\nu}\right)^{(\pm)} = -\sum_{\sigma=\pm}16\pi G^{(\sigma)}\left(T_{\mu\nu} - \frac{1}{3}\gamma_{\mu\nu}T\right)^{(\sigma)} \pm \frac{16\pi G^{(\pm)}}{3}\frac{\sinh(d/\ell)}{e^{\pm d/\ell}}\gamma_{\mu\nu}T^{(\pm)},$$

where the plus and minus refer to quantities on the wall with positive and with negative tension respectively. We thus derive

$$G^{(\pm)} = \frac{G_5\ell^{-1}e^{\pm d/\ell}}{2\sinh(d/l)}$$

giving us the shadow world/matter world Newton's constant in BD format. The induced metric is thus,

$$\bar{h}_{\mu\nu} = -16\pi G\frac{1}{\Box^{(4)}}\left(T_{\mu\nu} - \frac{1}{2}\gamma_{\mu\nu}T\right),$$

where

$$G = \ell^{-1}G_5$$

is the 4 Dimensional Newton's constant on our sheet. The motivation for such a model is as follows:

The vacuum energy density should contribute both to gravity and the Cosmological constant since

$$R_{\mu\nu} - \frac{1}{2}g_{\mu\nu}R - \Lambda g_{\mu\nu} = \frac{8\pi G}{c^4}T_{\mu\nu}$$

Observation tells us

$$|\rho_{vac}| < 10^{-29}\ g/cm^3 \sim 10^{-47}\ GeV^4 \sim 10^{-9}\ erg/cm^3$$

and

$$|\Lambda| < 10^{-56} cm^{-2}$$

The classical Higg's potential is given by

$$V(\phi) = V_0 - \mu^2\phi^2 + \lambda\phi^4 \quad \langle 0|\phi|0\rangle = v$$

With the known range of around 125 GeV based upon current high energy particle collisions. Where V is 250 GeV for the EW scale, we get when V=0

$$\rho_{vac}^{SSB} = -\frac{\mu^4}{4\lambda} \sim -\lambda v^4 \sim -10^5\ GeV^4 \quad \rho_{vac}^{OBS} \sim 10^{-47}\ GeV^4|$$

$$|\rho_{vac}^{SSB}| \sim 10^{56}\rho_{vac}^{OBS}.$$

This still leaves us a problem when it comes to the contribution of the quantum vacuum. The simplest solution is to propose a solution along similar lines to the so-called Higgless models(Kaplan and Sundrum 2005), but in this case two Higg's modes exist, one in the shadow world sheet and the other in our world sheet. You get the same cancellation of the quantum expectation value for the vacuum but in a Finite temperature metastable state that has a small cosmological constant.

We know from recent searches that our Higg's lies at around the125 GeV. The EW scale around 250 GeV as a primary is canceled by the shadow world sheet's negative Higg's of 125 leaving our Higg's value at 125 positive. This is in a metastable range which means further vacuum decay can take place. The EW scale itself is canceled by a shadow world sheet negative 250 GeV primary value which then removes any contribution via the EW scale to the quantum vacuum expectation value. The Quark-Gluon scale is itself canceled by a similar effect.

This leaves only the possible decay states of the Higg's as the origin of any cosmological constant. We know it's maximum value from observation and can further constrain it via the above equations to less that 10_{-5} GeV with observation further lowing it to less than 10_{-9} GeV. We then need a mechanism that keeps the metastable vacuum nearly stable with an increase in decay over time. But before that we will look at gravity a bit closer.

Starting with the Einstein equations(1) defined by

$$R_{MN} - \frac{1}{2}g_{MN}R = \frac{8\pi G}{c^4}T_{MN}.$$

In general relativity the gravitational force is represented by a Riemannian metric of curved space-time manifold M

$$\left(\frac{\partial}{\partial s}\right)^2 = g^{MN}(x)\frac{\partial}{\partial x^M} \otimes \frac{\partial}{\partial x^N}.$$

defined by the tensor product of two vector fields

$$E_A = E_A^M(x)\frac{\partial}{\partial x^M} \in \Gamma(TM)$$

where

$$\left(\frac{\partial}{\partial s}\right)^2 = \eta^{AB}E_A \otimes E_B.$$

The vector field

$$E_A \in \Gamma(TM)$$

are the smooth sections of tangent bundle

$$TM \to M$$

which are dual to the vector space

$$E^A = E_M^A(x)dx^M \in \Gamma(T^*M), \text{ i.e., } \langle E^A, E_B \rangle = \delta_B^A.$$

If we allow that that a spin-two graviton might arise as a composite of two spin-one vector fields(2) whereby we find the tensor product under the relationship

$(1 \otimes 1)_S = 2 \oplus 0.$

We then have the requirement of general covariance that gravity couples universally to all kinds of energy. Therefore the vacuum energy

$$\rho_{\text{vac}} \sim M_P^4$$

will induce a highly curved space-time whose curvature scale R would be

$$\sim M_P^2$$

we know that QFT is well-defined as ever in the presence of the vacuum energy because the background space=time still remains flat or at least behaves as if it was flat. So while any field for fundamental particles in Standard Model cannot be written as the tensor product of other two fields only composites can it would seem the most likely case for our composite would be two parts that unite and via addition of fields cancel part of

$$\rho_{\text{vac}} \sim M_P^4$$

The before mentioned mechanism solves that cancellation problem. But we still then have to find the two part carrier. One could propose gravitons and anti-gravitons or one can think in terms of graviphotons. In either case you have a quantum carrier for gravity. However, the zero-trace quadrupole moment tensor of a system of charges is defined as

$$\nabla^2 V - \frac{1}{c^2}\frac{\partial^2 V}{\partial t^2} = 4\pi G\rho$$

The scalar potential is given by

$$V = -N\left(i\left(\frac{1}{kr} - \frac{3}{(kr)^3}\right) - \frac{3}{(kr)^2}\right)\left(3\cos^2\theta - 1\right)e^{i(kr-\omega t)}$$

With the following relations in Maxwell format

$$B = \frac{\omega}{c^2}(r \times \nabla V) \qquad\qquad E = \frac{ic^2}{\omega}(\nabla \times B)$$

By substituting the relations the value of n can be determined
$\varepsilon_0 = -1/(4\pi G)$

the result yields

$$B_\phi = \frac{6N\omega}{c^2}\left[\left(\frac{1}{kr} - \frac{3}{(kr)^3}\right)i - \frac{3}{(kr)^2}\right]\left[Cos(\theta)Sin(\theta)\right]e^{i(kr-\omega t)}$$

$$E_r = 6Nk\left[\left(\frac{-3}{(kr)^3}\right)i - \frac{1}{(kr)^2} + \frac{3}{(kr)^4}\right]\left[3Cos^2(\theta)-1\right]e^{i(kr-\omega t)}$$

$$E_\theta = 6Nk\left[\left(\left(\frac{1}{kr}\right) - \frac{6}{(kr)^3}\right)i + \frac{6}{(kr)^4} - \frac{3}{(kr)^2}\right]\left[Cos(\theta)Sin(\theta)\right]e^{i(kr-\omega t)}$$

where: $N = -G\,m\,s^2\,k^3$, G = Grav const., m = mass, s = Dipole length, k = Wave number

in the limit

(kr → 0).

Now if we postulate that this field is itself split into two components then the resultant is a simple dipole field with conventional EM Maxwell equations. One set of equation would involve negative energy and the other positive energy. This offers a way to test this idea out involving a search for an em signal from a predictable origin point where quadrapole radiation should be generated at about half the amplitude to be expected of the normal quadrapole gravity wave detection method, but at the same frequency of the expected quadrapole gravity field.

We shall then return to the origin of the cosmological constant and the issue of the metastable state of the vacuum. One area we should be looking for is a decay of the Higg's into two photons of 31 to 32 GeV with a missing mass in the event of around 62 to 63 GeV. The two photons would be at the upper end of QED stability and as such, when combined with the extra missing mass term result in an unballance of the canceling of the quantum vacuum's expectation value resulting in an increase of vacuum pressure over time. This vacuum pressure changes results in an evolving cosmological constant.

The missing mass would be the long sought dark matter formed out of an actual stable minimum that is not coupled to the EW scale and as such does not radiate. Such a free stable Higg's boson would have only one means by which we can detect it, it's gravity signature.

The RS model, which this model is based upon, circumvents the need of compactifying all but the three observed spatial dimensions by including a bound state of the massless graviton on the brane(1). At distances defined by

$r \gg L,$

the model implies a correction to the Newtonian gravitational potential of a body of mass M. This correction is given by

$$U_{\mathrm{RS}} = -k\frac{GM}{r}\left(\frac{\mathrm{L}}{r}\right)^2,$$

where G is the Newtonian constant of gravitation, and k can assume different values depending on the schemes of regularization adopted(2). This value can be derived out of the before mentioned Newtonian formula. The local and global canceling of the ZPF modes tends to constrain this itself, except in the case of decayed Higg's states with the uncoupling of the cancelation effect so that Dark Matter in effect should display the value of K, thus displaying a slight correction to the standard Newtonian. It is through this avenue that the effects of the other world sheet can be detected.

In essence, our current attempts at detecting via direct measurement, both gravity waves and the graviton predicted by theory are very possibly doomed from the start simply because we may actually be trying to measure a sort of AC current via DC methods, in a manor of speaking. The problem is we can only measure one part that is within our observation arena, when in fact, the full other half of the signal is outside of that directly observable arena.

This is where Science sometimes runs into limits on our observational ability and one of the prime reasons it can be argued from a strict scientific method that ruling out the existence of something like a Creator may itself be beyond what Science can directly study implying that any attempt to use Science to prove or disprove God is simply based upon a false assumption to begin with. One when dealing with Science must keep in mind the old adage about assumption makes an ass out of you and me.

References

1.) L. Randall and R. Sundrum, Phys. Rev. Lett. 83, 4690 (1999).

2.) E. Jung, S. H. Kim, and D. K. Park, Nucl. Phys. B 669, 306 (2003).

A Relook at the No-Boundry Idea

In multiverse modeling our universe arose as a decaying false vacuum state out of a much larger false vacuum frame. As such the real boundary of our universe would be the domain wall that formed between us and that false vacuum state. But the question arises how does that wall and the external false vacuum state and any other possible universes effect us.

Starting with

$$\left(ds^2\right)^+ = g^+_{\mu\nu}dx^\mu_+ dx^\nu_+ = -\left(1 - \frac{2M}{r}\right)dt^2 +$$

$$\frac{dr^2}{1 - \frac{2M}{r}} + r^2\left(d\theta^2 + sin^2\theta\, d\phi^2\right).$$

and assuming

$$\Delta t = \int_{R_s}^{R_0} \frac{dr}{1 - \frac{2M}{r}} \to \infty,$$

as the Schwarzschild time interval we discover that while the outer domain wall effects evolution internal within our space-time, external it has no effect on any sort of evolution of the false vacuum frame itself in which it is embedded. Especially when we consider and assume that

$$a\left(\eta\right)$$

the scale factor for that frame and that as such the time factor for it is zero or even if it evolves such a time factor would be different from our own.

But this changes in a quantum perspective. The horizon radius in respect to the false vacuum domain

$$R = R_s + \delta R_s$$

has an uncertainty expressed as

$$\delta R_s$$

yielding a time interval of

$$\int_{t_f}^{t_i} \frac{d\eta}{a^2(\eta)} = \int_{R_s+\delta R_s}^{R_0} \frac{dr}{1 - \frac{R_s}{r}}$$

$$\sim R_s \ln \frac{R_0 - R_s}{\delta R_s}.$$

Thus, the near infinite uncertainty implies at a quantum scale some sort of entangled effect not only having effect at that scale on the false vacuum, but also upon other possible existing domains themselves embedded in such a frame.

Following the more simple model of the Nambu-GoTo we get an action expressed as

$$S = \int d^4 x \sqrt{-g} \left[-\frac{1}{16\pi} R + \frac{1}{2} (\partial_\mu \Phi)^2 \right]$$

$$- \sigma \int d^3 \zeta \sqrt{-\zeta} + S_{obs}.$$

which includes the

$$\zeta^a$$

as denoting the coordinates of the wall itself expressed in (1+2) dimensional format. Following an Israel format for the junction conditions we can graph such as:

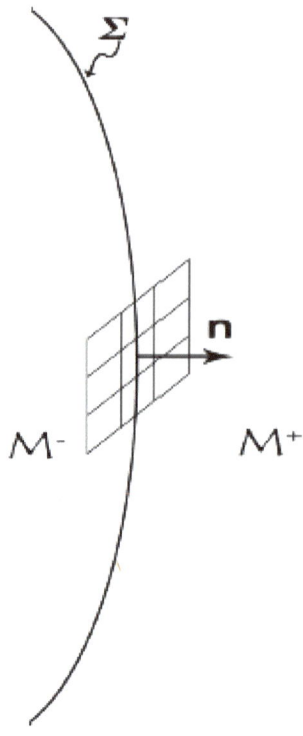

FIG. 1: The timelike hypersurface Σ is shown which forms the boundary of the individual spacetime regions M^- and M^+. The normal $\mathbf{n}$ is also depicted.

In this case we are the M- and the false vacuum state is the M+ region. If we then take

$$R^\alpha{}_{\beta\mu\nu} = R^\lambda{}_{\gamma\sigma\rho}h^\alpha{}_\lambda h^\gamma{}_\beta h^\rho{}_\nu +$$
$$\epsilon\left(K^\alpha{}_\mu K_{\beta\nu} - K^\alpha{}_\nu K_{\beta\mu}\right),$$

and

$$\nabla_\alpha K_{\beta\mu} - {}^{(3)}\nabla_\mu K_{\beta\alpha} = R^\lambda{}_{\sigma\rho\delta}n_\lambda h^\sigma{}_\beta h^\rho{}_\alpha h^\delta{}_\mu,$$

we get the relationship of interior to exterior both as

$$E_{\mu\nu}n^\mu n^\nu = -\frac{1}{2}\epsilon^{(3)}R + \frac{1}{2}\left(K^2 - K_{\alpha\beta}K^{\alpha\beta}\right),$$

$$E_{\mu\nu} h^\mu{}_\alpha n^\nu = -\left({}^{(3)}\nabla_\mu K^\mu{}_\alpha - {}^{(3)}\nabla_\alpha K \right).$$

The Einstein field equation becomes

$$E_{\mu\nu}^\pm = \kappa T_{\mu\nu}^\pm,$$

where

$$R_{\mu\nu}$$

is the Ricci curvature tensor,

$$R = g^{\mu\nu} R_{\mu\nu}$$

is the Ricci scalar and

$$T_{\mu\nu}$$

is the stress energy tensor. We now derive that

$$E_{\mu\nu} = R_{\mu\nu} - \tfrac{1}{2} R g_{\mu\nu}$$

showing that there exists an evolution entanglement itself with the domain wall and both frames. Our region interior of this wall takes the metric form

$$\left(ds^2\right)^- = g_{\mu\nu}^- = -dT^2 + a^2\left(\eta\right)\Big[dr^2 + r^2\left(d\theta^2 + \sin^2\theta d\phi^2\right)\Big].$$

If mass vanishes at the wall the

$$a\left(\eta\right)$$

becomes suppressed. If not then it comes into play. If we then (1+1) slice of the Interior

$$\alpha = \frac{dT}{d\tau} = \sqrt{1 + a^2 R_\tau^2},$$

with the exterior

$$\beta = \frac{d\eta}{d\tau} = \frac{1}{1 - \frac{2M}{R}} \sqrt{1 - \frac{2M}{R} + a^2 R_\tau^2},$$

We get a time relationship of

$$\dot{T} = \frac{dT}{d\eta} = \sqrt{1 - \frac{2M}{R} - \frac{\frac{2M}{R} a^2 \dot{R}^2}{1 - \frac{2M}{R}}},$$

which shows a relationship with

η.

Since the domain wall is located at

$$u \rightarrow -\infty$$

with respect to us the question arises what effect would its shift outward or inward have on our present observation of the cosmos. Only if that horizon where to cross our current position along

$$u \rightarrow -\infty$$

would we ever see it's effect. And given the long finite time for any signal to reach us any outward expansion would itself take an equally long time to arrive at our position. Now if we already crossed such a boundary where those signals have reached us we might detect in the last case an accelerated expansion of the cosmos. And in the first case we would not be here to even reflect upon that question if the boundary is already here and if only its signal has arrived all we would detect is a recollapse of the cosmos.

The basic point in all this is the idea of a No-Boundary may perhaps need to be reexamined. In General, under the Multiverse idea our space-time arose from a huge false vacuum sea as a sort of bubble. As such, given that the false vacuum sea would still exist in its own right there should be a boundary or cauchy surface out there. This Boundary is a domain wall that seperates one energy state(ours) from the other energy state(the False Vacuum). That Boundary also has the effect of forcing our space-time out towards infinity to appear flat even if its overall mass energy density would require another condition. In that aspect the actual Boundary has an important role in the evolution of the Cosmos around us via control of the overall geometry of space-time itself.

In this aspect, the boundary becomes actually a pre-built in geometry to the cosmos in which we live. That geometry being a predisposition towards a spatially flat appearing cosmos who's over all mass/energy density may or may not appear to support such a state. In fact, this Prior geometry has an influence on the overall history of the cosmos when you take into consideration that even with exotic energy causing an accelerated expansion, such an expansion will appear to go on in a nearly flat space-time.

What you encounter here is that even though from our local perspective the Cosmos may appear infinite, knowing that it has only been in existence for around 13 billion years tells us the Cosmos we live in is not infinite at all. We then try by observation and theory to predict certain aspects that govern the overall shape of space-time itself. Some of those aspects would in themselves argue the Universe should be spatially closed, some evidence would argue it should be an open model. Yet, the truth is we live in a Cosmos that tends to defy the common sense evidence in favor of a possibly eternally expanding, possibly closed flat space-time that we by theory would expect to be one thing when in fact it is another.

Now if we really look at this boundary in effect it is as if our Cosmos was inside of a Schwarzschild sphere. If we follow that to its conclusion then in essence the overall fate of our Cosmos depends a lot upon if there is Hawkin Radiation from its surface or event horizon back into the false vacuum sea. This is turn is somewhat governed by the relationship between dark matter and dark energy. If the amount of dark matter from the decay of Higg's Bosons over time results in an overall mass/energy density that can overcome the dark energy and that forced nearly flat state it is possible for the Cosmos to eventually evolve into a closed, collapsing nearly flat space-time at least till it reaches some super critical size at which it fully collapses back into the false vacuum sea via such Hawkin radiation.

Reference

Math treatment comes from recent work by

Jackson Levi Said

And

Kristian Zarb Adami

In

Cosmological Effects on Black Hole Formation

Some Interesting Pictures and Graphs

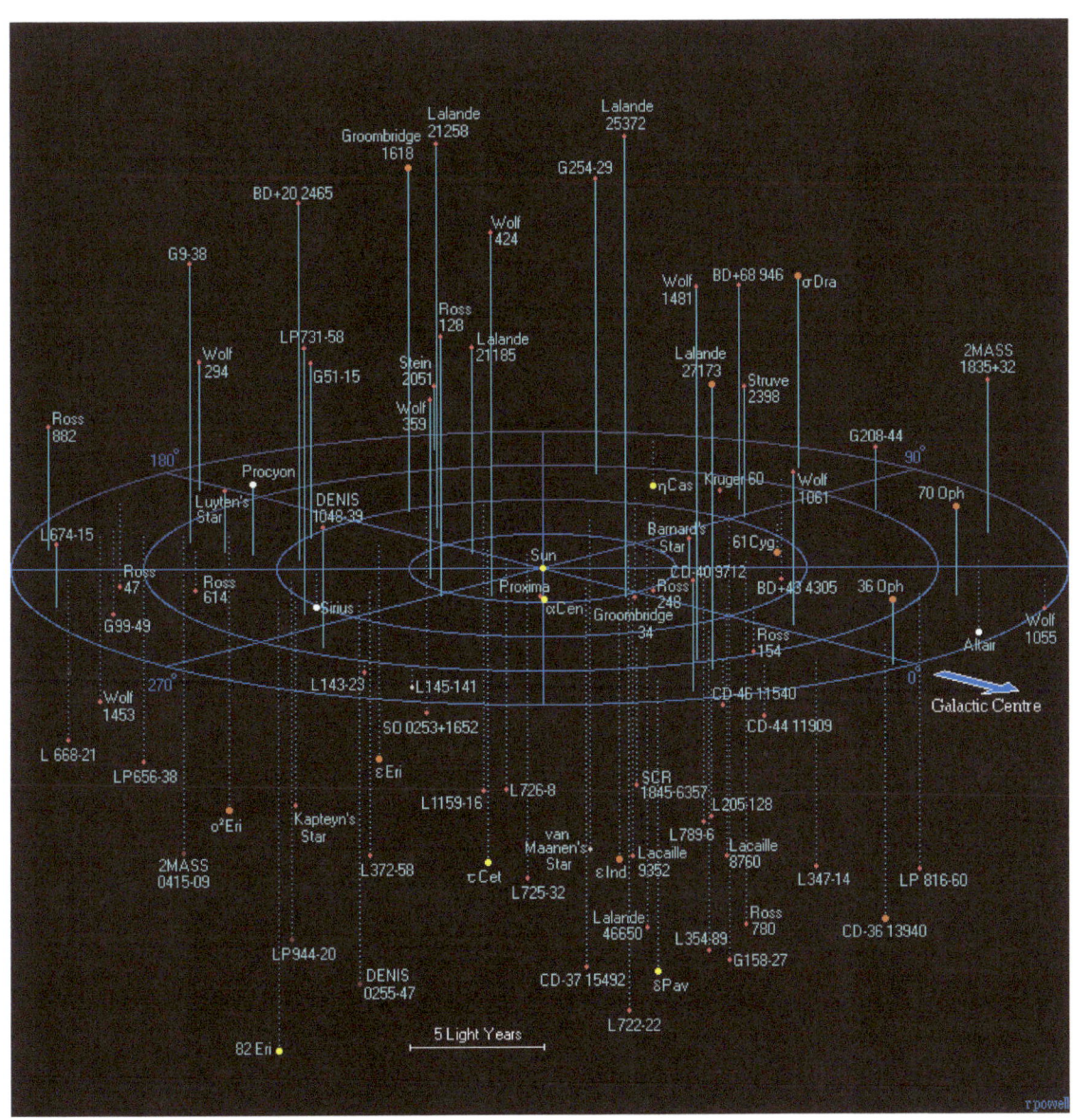

Local area out to 20 Light Years.

Artist Conception of a Blackhole

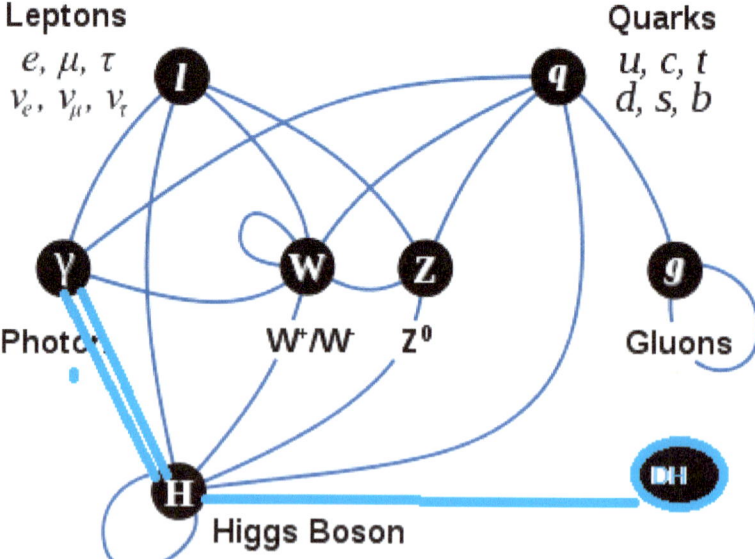

A possible newer Higgs Boson decay chart

Picture of another galaxy similar to our own.

Area in our Galaxy of Star Birth.

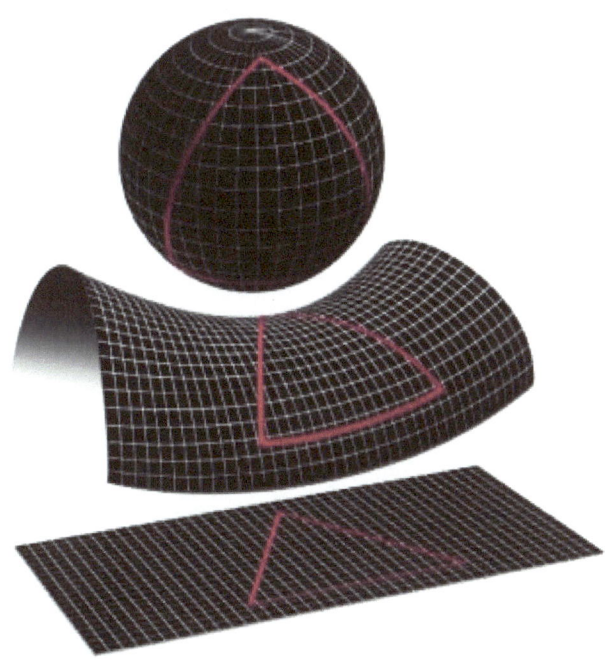

Different possible shapes of the Cosmos.

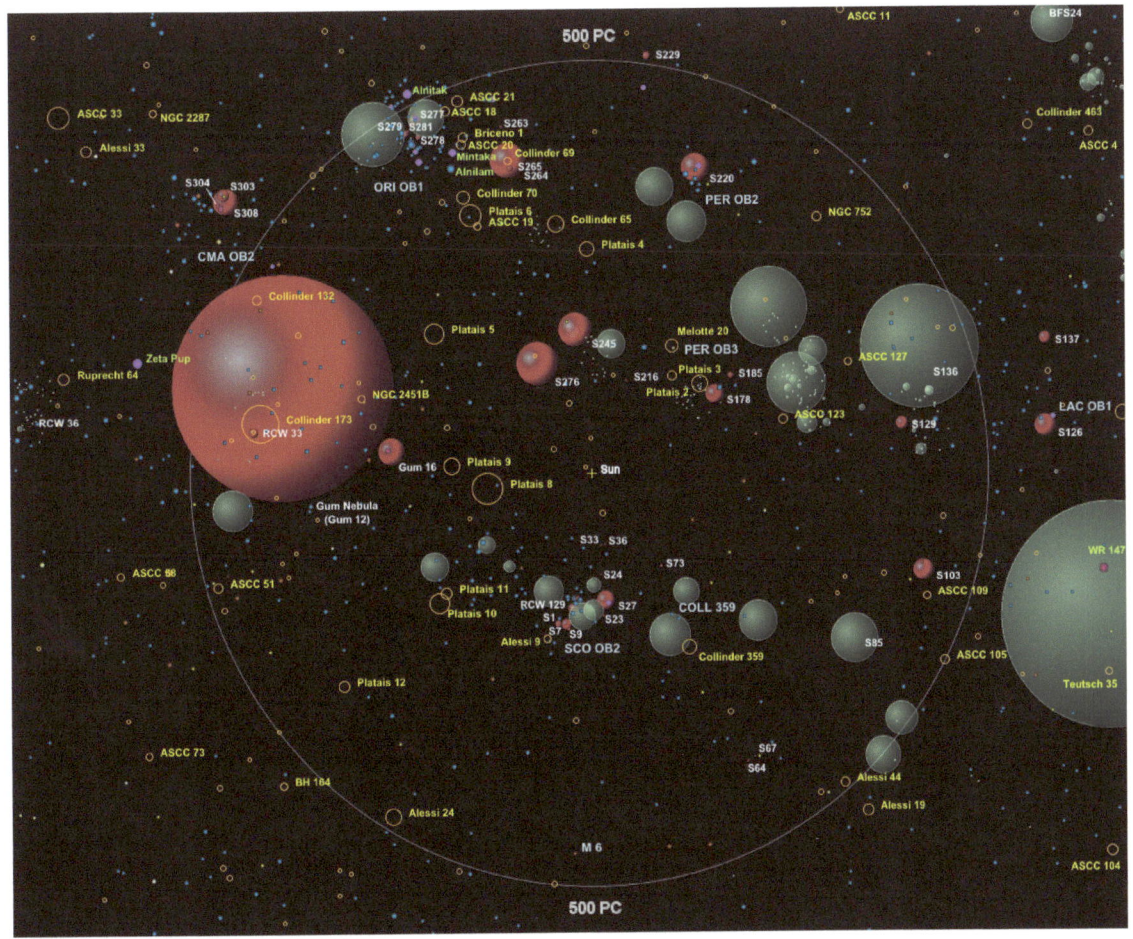

Star Cluster out to 500 PC.

Cross Section of our Galaxy.

The Andromeda Galaxy.

Orion Cloud.

Illustration of Brown Dwarfs in local area.

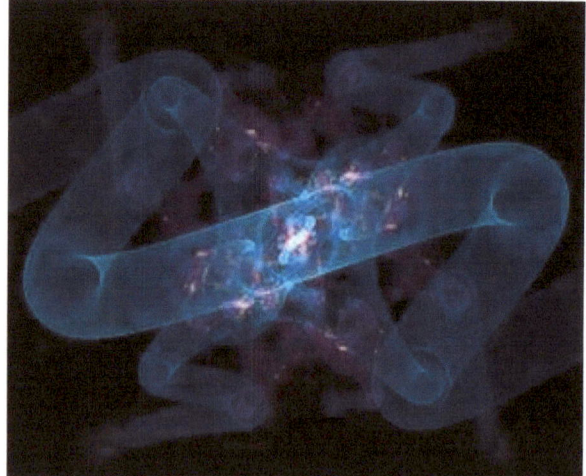

Artists illustration of wormhole like
Structure of space-time at quantum
Scales.